SCIENTIFIC MENTAL HEALING

Published @ 2017 Trieste Publishing Pty Ltd

ISBN 9780649114047

Scientific mental healing by H. Addington Bruce

Except for use in any review, the reproduction or utilisation of this work in whole or in part in any form by any electronic, mechanical or other means, now known or hereafter invented, including xerography, photocopying and recording, or in any information storage or retrieval system, is forbidden without the permission of the publisher, Trieste Publishing Pty Ltd, PO Box 1576 Collingwood, Victoria 3066 Australia.

All rights reserved.

Edited by Trieste Publishing Pty Ltd.
Cover @ 2017

This book is sold subject to the condition that it shall not, by way of trade or otherwise, be lent, re-sold, hired out, or otherwise circulated without the publisher's prior consent in any form or binding or cover other than that in which it is published and without a similar condition including this condition being imposed on the subsequent purchaser.

www.triestepublishing.com

H. ADDINGTON BRUCE

SCIENTIFIC MENTAL HEALING

SCIENTIFIC MENTAL HEALING

BY
H. ADDINGTON BRUCE

AUTHOR OF "THE RIDDLE OF PERSONALITY," ETC.

BOSTON
LITTLE, BROWN, AND COMPANY
1911

Copyright, 1911
BY LITTLE, BROWN, AND COMPANY.

All rights reserved

Published, September, 1911

Printers
S. J. PARKHILL & CO., BOSTON, U.S.A.

PREFACE

THE chief aim of the present volume is to provide the general reader with a brief, yet it is hoped sufficiently comprehensive, account of the principles underlying scientific psychotherapy; and to afford some idea of the methods by which it is applied in the treatment of disease, and also of the maladies to which it is applicable. For this reason the use of technical terms has been avoided as far as possible, and there has been a liberal citation of illustrative cases for the purpose of bringing the principles and the methods concretely to the reader's mind, and in order to emphasize the fundamental differences between psychotherapy of the scientific type and the psychotherapy of " faith healing."

The book is thus of the nature of a " popular" manual, and may perhaps be described as a primer in scientific psychotherapy. But the writer trusts that it will not on that account be found devoid of usefulness to the physician and psychologist, and that it may be the means of stimulating in some measure a broader interest in investigations that are unquestionably of tremendous importance to humanity, particularly in this age of hurry, unrest, and "nerve strain." There can be no doubt that functional mental and nervous diseases, as well as the true insanities, are increasing in civilized countries; and statistics such as those gathered, for instance, by the United States Census Bureau, would seem to indicate that they are increasing most rapidly in the countries of highest economic development. Scientific mental healing affords a means of coping with this growing evil, as it will be one of the writer's main objects to make clear.

The three essays, " Psychology and Everyday Life," "Half a Century of Psychical Research," and "William James" are included because, while they relate only indirectly to the subject of mental healing, they contain information bearing on it in several important respects. Thus, the "Psychology and Everyday Life," though primarily concerned in setting forth possible applications of the results of psychological experimentation in other than a medical way, also surveys methods of mental analysis that are helpful to the physician; the "Half a Century of Psychical Research" emphasizes the debt which modern medical psychology owes to the scientific investigation of the phenomena of spiritism; and, finally, the essay on the late Professor James is included by way of appreciation of the notable services rendered by this eminent American psychologist to both medical psychology and psychical research.

In preparing the various essays, aid has been sought from and generously rendered by leaders in psychological and psychopathological investigation, and to these gentlemen grateful acknowledgment should be made. The writer's thanks are also due the editors of *The Outlook, The American Magazine, The Cosmopolitan Magazine,* and other publications in which the essays originally appeared. Each essay, it should perhaps be added, has been revised for present publication.

H. ADDINGTON BRUCE.

CAMBRIDGE, MASS., June, 1911.

CONTENTS

		PAGE
	PREFACE	iii
I	THE EVOLUTION OF MENTAL HEALING	1
II	PRINCIPLES AND METHODS	39
III	MASTERS OF THE MIND	66
IV	HYPNOTISM AS A THERAPEUTIC RESOURCE	102
V	SECONDARY SELVES	124
VI	PSYCHOLOGY AND EVERYDAY LIFE	156
VII	HALF A CENTURY OF PSYCHICAL RESEARCH	194
VIII	WILLIAM JAMES — AN APPRECIATION	230
	INDEX	253

SCIENTIFIC MENTAL HEALING

I

The Evolution of Mental Healing

MENTAL healing — or psychotherapy, to give it its technical name — is to-day practiced on a most extensive scale and under several forms, some primarily religious in character, others based wholly on the results of scientific investigation. Each particular system, whether religious or scientific, possesses characteristics peculiar to itself and marking it off more or less sharply from every other system. But all have this in common, that they rest at bottom on two general principles — the power of the mind over the body, and the importance of suggestion as a factor in the cure of disease.

Moreover, all have a common ancestry, dating back directly to the closing years of the eighteenth century, and indirectly to the dim ages of antiquity.

Psychotherapy, indeed, might well be cited in support of the old adage that there is nothing new but what has been forgotten. Traces of it are to be found almost as far back as authentic history extends, and even allusions to methods which bear a strong resemblance to those of modern times. The literature and monumental remains of ancient Egypt, Greece, Rome, Persia, India, and China reveal a widespread knowledge of hypnotism and its therapeutic value. There is in the British Museum a bas-relief from Thebes which has been interpreted as representing a physician hypnotizing a patient by making "passes" over him. According to the Ebers papyrus, the "laying on of hands" formed a prominent feature of Egyptian medical practice as early as 1552 B. C., or nearly thirty-five hundred years ago; and it is known that a similar mode of treatment was employed by priests of Chaldea in ministering to the sick.

So, also, the priests of the famous Temples of Health are credited with having worked numerous cures by the mere touch of their hands. In connection with these same Temples of Health were sleeping-chambers, repose in which was supposed to be exceptionally beneficial. A learned man of Bithynia, who won considerable fame at Rome as a physician, systematically made use of the "induced trance" in the treatment of certain diseases. Plautus, Martial, and Seneca refer in their writings to some mysterious process of manipulation which had the same effect — that is, of putting persons into an artificial sleep. And Solon sang, apparently of some form of mesmeric cure:

"The smallest hurts sometimes increase and rage
More than all art of physic can assuage;
Sometimes the fury of the worst disease
The hand, by gentle stroking, will appease."

Many other instances might be mentioned testifying to the remarkable extent to which psychotherapy, in one form or another, was utilized in the countries of the ancient world.[1] This, of course, does not necessarily imply that the ancients had any real understanding of the psychological and physiological principles governing its operation. On the contrary, there is every reason to believe that they used it much as do too many of the mental healers of to-day — on the basis of a "faith cure" pure and simple, with no attempt at diagnosis, and in hit-or-miss fashion. It was not until the very end of the Middle Ages, so far as history informs us, that anything even remotely resembling a scientific inquiry into its nature and possibilities was undertaken, and then only in a faint, vague, indefinite way, by men who were metaphysicians and mystics rather than scientists.

The first of these, Petrus Pomponatius, a sixteenth-century philosopher, sought to

[1] The reader who wishes to study this phase of the subject in more detail will find much curious information in J. C. Colquhoun's "Isis Revelata;" and, among later works, in R. M. Lawrence's "Primitive Psychotherapy and Quackery."

prove that disease was curable without drugs, by means of the "magnetism" existing in certain specially gifted individuals. "When those who are endowed with this faculty," he affirmed, "operate by employing the force of the imagination and the will, this force affects their blood and their spirits, which produce the intended effects by means of an evaporation thrown outwards." Following Pomponatius, John Baptist van Helmont, to whom medical science unquestionably owes a great deal, also proclaimed the curative virtue of magnetism, which he described as an invisible fluid called forth and directed by the power of the human will. Other writers, notably Sir Kenelm Digby, William Maxwell, and the Rosicrucian Robert Fludd, advanced the same ideas; and at least one of them, Sir Kenelm Digby, laid stress on the power of imagination as an agent in the cause as well as the cure of disease, compiling, in a curious little treatise published in 1658, and before me as I write, as interesting a collection of illustrative cases as is contained in many books dealing with modern psychotherapy.[1]

For various reasons, however, these early theorists failed to gain the confidence of, or even a hearing from, their contemporaries. As an enthusiastic advocate of the claims of the magnetic " school " of mental healing has explained in painstaking detail, " the style in which most of their treatises were written was so shrouded in mystical expression; the vague and unsatisfactory theories in which their authors delighted to indulge tended so much to obscure the few facts which they really developed; and the opinions they announced were so much at variance with the common philosophical systems, as well as with the ordinary experience of life, that no attempts appear to have been made to ascertain the truth or falsehood of their principles by a fair appeal to the decisive test of scientific experiment. About

---

[a] Sir Kenelm Digby's "A Late Discourse made in a Solemn Assembly of Nobles and Learned Men at Montpelier in France, Touching the Cure of Wounds by the Powder of Sympathy."

that period, too, chemical science, and its application to medicine, began to be cultivated with great zeal and prosecuted with eminent success, and it was not to be expected that much attention should be devoted to a subject so remote from the fashionable pursuits of the age." In fact, all classes united in condemning the magnetists as crack-brained visionaries no better than the astrologers and alchemists. It remained for an inquirer of a far later generation, burrowing among the dust of a university library, to glean from their forgotten volumes the clues necessary to enable him to rediscover psychotherapy and present its marvels to a wondering world.

This was Franz Anton Mesmer, the first and by far the most picturesque of the long line of modern mental healers. A native of Switzerland, where he was born in 1734, Mesmer removed in early manhood to Vienna for the purpose of studying medicine. Incidentally he also became deeply interested in the study of astrology and other occult subjects, and this led him to make the acquaintance of the writings of Pomponatius, van Helmont, and their fellow-mystics. The notion that there existed in nature a magnetic force which might be utilized for therapeutic purposes made a strong appeal to his always exuberant imagination, and shortly after receiving his doctor's degree he began some experiments intended to prove or disprove the existence of such a force. He found it possible, according to a public statement which he made in 1773, to cure many maladies merely by the application of an iron rod to the body of the patient; and he further declared — almost precisely as van Helmont had affirmed long before — that the healing agent was an impalpable fluid emanating from his own person and conveyed to the patient by means of the iron rod. To this impalpable but seemingly all-powerful fluid Mesmer gave the name of "animal magnetism."

His confident belief that he had made a discovery which would revolutionize the science of medicine was not shared by his professional brethren. Nor was their skepticism lessened when they came to inquire closely into his methods. In order to accommodate the greatest possible number of patients, Mesmer had invented a "magnetic tub," a large circular vat containing various chemicals and covered with a lid pierced with holes through which passed iron rods. About this tub the patients grouped themselves, each taking hold of one of the iron rods. When all was in readiness Mesmer would enter, clad in a lavender-colored robe and carrying a small metallic wand. No one spoke, and the silence of the room was broken only by the soft strains of distant music. The "magician," as his critics angrily styled Mesmer, would then gaze intently at the patients, and, striding majestically up to them, touch each with his wand.

At the touch some would burst into hysterical laughter, others into tears, and others again would fall into convulsions, finally lapsing into a state of complete insensibility. It could not have been an edifying sight, and undoubtedly there was a large strain of the charlatan in this pioneer psychotherapist. But it also seemed certain that he was effecting some remarkable cures with his "magnetic tub," and, while orthodox physicians scoffed and sneered, the sick flocked to him for relief. Especially was this the case after 1778, when Mesmer left Vienna to make his home in Paris. There he was fortunate enough at the outset to win an influential convert in the medical adviser to the Count d'Artois, and to find favor with the fashionable world. Such was the interest he excited that in March, 1781, the King offered him a pension of thirty thousand livres on condition that he made public the secret of his treatment. Mesmer rejected this offer, but two years later opened a school for the instruction of suitable pupils in animal magnetism, or mesmerism, as it was beginning to be called.

He could have devised no better means for

propagating his views and keeping the subject prominently before the public. His pupils traveled far and wide, giving mesmeric treatment and instructing others in the new science. Interest was also heightened by the fact that fresh discoveries were constantly being announced. The Marquis de Puységur, one of Mesmer's earliest disciples, found to his astonishment that mesmerized subjects sometimes fell into a profound sleep, during which they would respond intelligently to questions put to them by the mesmerist and obey his slightest command, but on awaking be quite unaware of what had transpired. This was the first intimation in modern times of the phenomena of the "induced trance," so common to-day in hypnotic practice. It was also found — greatly to the relief of timid folk — that the good effects of mesmerism might be obtained without going through the preliminary stages of hysteria and convulsions, and that it was possible for the operator to communicate the salutary magnetic fluid simply by gazing into the eyes of his patient and gently stroking his face.

Thus it came about that, although Mesmer himself eventually fell into disrepute and died in obscurity, mesmerism took root and flourished not only in France but in almost every other European country, and more particularly in Germany, Switzerland, Denmark, and Russia. The French Revolution and the Napoleonic wars for a time checked its practice, but with the restoration of peace its exponents again sprang into widespread activity and popularity.[1] At the same time serious efforts began to be made to find an adequate explanation for its singular phenomena. While the fluidic theory was still upheld by a great majority, there were a few investigators discerning enough to perceive what is now almost universally recognized — namely, that the actual motive force was nothing more or less than sug-

[1] Much interesting information about Mesmer and mesmerism will be found in the first volume of Frank Podmore's "Modern Spiritualism," which also gives helpful bibliographical references. Consult also the same writer's "Mesmerism and Christian Science."

gestion. In 1815 the Abbé Faria, a learned Portuguese, demonstrated experimentally that the hypothesis of a magnetic fluid was quite superfluous; that in order to induce the mesmeric state it was only necessary to provoke a high state of expectancy in the patient. The cause of the trance, he said, was not in the operator but in the person to be entranced — was, in other words, purely subjective. Not long afterwards a brilliant young Frenchman, Alexandre Bertrand, voiced the same conclusion in a work that has become a classic in the literature of mental healing.

Unfortunately, the time was not ripe for medical science to appreciate and profit by the truth thus brought to light. Physicians in general still kept rigidly aloof from mesmerism, denouncing its practitioners as impostors and its devotees as fools — an attitude for which they felt they had ample justification in the notorious circumstance that many of the later mesmerists were mere showmen, reaping a golden harvest by exhibiting their entranced subjects on the public platform. Men of conservative and serious mind were still further repelled by a growing tendency to attribute the more striking phenomena of the trance to some supernatural influence. Under such conditions it is not surprising that the findings of Faria and Bertrand were completely ignored, and that it was many years before a real beginning was made to scientific psychotherapy.

Even then this was largely due to chance. In 1841 a French mesmerist visited the English city of Manchester and gave a number of demonstrations that won for him an enthusiastic following and enormous audiences. A local physician, Dr. James Braid, who shared to the full the prevailing belief of the educated classes that mesmerism was compounded almost equally of deception and credulity, felt it his duty in the interest of the public good to investigate and expose the Frenchman's "tricks." To his surprise he found himself obliged to admit that, whatever their explanation, the mesmeric phenomena were unquestionably genuine. He resolved to continue his investigations. One fact which particularly impressed him was the inability of the mesmerized subjects to open their eyes. Attributing this to exhaustion of the optic nerve, and shrewdly guessing that the mesmeric trance resulted from modifications of the nervous system, he succeeded in proving that it was possible for persons to mesmerize themselves by gazing fixedly at some small and bright object, provided only that while so gazing they concentrated their thought as well as their vision, putting themselves into a state of "expectant attention." He also demonstrated by hundreds of experiments that persons so entranced were peculiarly suggestible, and that this condition of extreme suggestibility was sufficient to account for their ready obedience to the commands of the mesmerist.

To put it otherwise, Braid, like Faria and Bertrand before him, had hit upon the master fact of psychotherapy — suggestion. What was no less important, he had cleared the air by showing that it was not at all necessary to resort to the assumption of any such force as a magnetic fluid, mesmeric influence, or other unknown and mysterious agency. To distinguish his system from that of the mesmerists, he invented the term hypnotism and at once began to utilize hypnotic therapeutics as an auxiliary in the treatment of disease, hypnotizing those of his patients who would allow him to do so, and while they were in the hypnotic state making curative suggestions to them. For some time, however, there was scarcely a physician venturesome enough to follow his example. The old prejudices were hard to down, and medical opinion deemed Braid little better than the mesmerists.

In fact, it was not until 1860, the year of his death, that a successor appeared in the person of a Frenchman, Dr. A. A. Liébeault, to confirm and surpass the results Braid had obtained, and, by persistent, tireless endeavor, gradually compel recognition of the therapeutic helpfulness of suggestion as applied in the hypnotic trance. After long and careful experimentation, Liébeault, who had begun his career as a struggling country doctor, opened a public dispensary in the town of Nancy, and announced that he would treat free of charge all who would submit to be hypnotized. At first patients came timidly enough, and in small numbers. But so soon as it was discovered that hypnotism as administered by him hurt nobody and benefited many, his rooms were thronged with eager applicants.

His method of treatment was in sharp contrast with the sensational procedure of Mesmer and Mesmer's disciples. After a medical examination to determine the exact nature of the disease, the patient would be asked to sit in an armchair, make himself as comfortable as possible, and " think of nothing at all." Liébeault, speaking in an even, monotonous tone, would then inform him that his eyes would soon begin to feel heavy, that he could no longer keep them open, and that he would soon be sound asleep. Repeating this assurance firmly and authoritatively, the eyes of the patient would close, and he would pass into the hypnotic state, seemingly quite unconscious, but in reality alert to every word spoken by Liébeault, who would ply him with suggestions appropriate to his case. If he had been suffering from insomnia, Liébeault would assure him that he would henceforth sleep well; if he were a victim of neuralgia, the promise was made that the pain would disappear; and similarly with all manner of maladies.

Liébeault was not always successful; he found some patients whom he could not hypnotize, and others whom he failed to benefit. But he was successful in so many instances that he became widely talked about as a modern worker of miracles. He himself protested vigorously that there was nothing miraculous in his cures. " It is all a matter of suggestion," he would say. " My patients are suggested to sleep, and their ills are suggested out of them. It is very simple, once you understand the laws of suggestion." Other physicians, hitherto skeptical, began to betray a desire to learn something of these wonderful laws. First one, then another, made his way to Liébeault's humble dispensary. From all parts of France inquirers came, and presently from foreign countries — from England, Germany, Austria, Russia, Norway, Sweden, Denmark, Holland, Belgium, Italy, and even distant America. Elsewhere independent investigations were set on foot — notably at the Salpêtrière, in Paris, under the leadership of the celebrated Dr. Charcot. A new era had dawned for hypnotism, and the foundations of the scientific psychotherapy of the present day had at last been securely laid.[1]

Meantime another kind of psychotherapy had been in process of evolution. This was the religious psychotherapy now so well known under its principal forms of Christian Science and the New Thought. It,

[1] In "The Riddle of Personality," by the present writer, will be found a fairly comprehensive bibliography of the literature dealing with both early and late hypnotic investigation.

too, is an outgrowth of mesmerism, and is of special interest to us as being a distinctly American development, tracing its beginnings to 1836, when mesmerism was introduced into the United States by a young Frenchman, Charles Poyen, who had settled in New England the previous year.

As had been the case abroad, the phenomena of the trance condition appealed strongly to the popular imagination, but scarcely at all to men of science. It was generally believed that, even granting their genuineness, no useful purpose would be served by investigating them, and scientific curiosity was also chilled by the clamorous insistence with which sundry pseudo-scientists advanced all sorts of fantastic theories as " explanations " of the singular influence exercised by mesmerists over their sleeping subjects.

Thus Dr. J. S. Grimes, a professor of medical jurisprudence and a dabbler in phrenology, suggested that it was due to the action of an atmospheric force which he called etherium. Dr. J. R. Buchanan, another phrenologist, preferred the hypothesis of a subtle emanation from the nervous system. A clerical investigator, the Rev. J. B. Dods, sought to explain it on an electrical basis. Among all the "authorities" who, in the decade 1840-50, bombarded the American public with their quaint ideas and quarreled violently with one another, only one, the Rev. Laroy Sunderland, seems to have had the least glimmering of the truth. "When," declared Sunderland, "a relation is once established between an operator (or any given substance, real or imaginary, as the agent) and his patient, corresponding changes may be induced in the nervous system of the latter (awake or entranced) by mere volition, and by suggestions addressed to either of the external senses."[1] Had he not made the mistake of upbuilding on this foundation a mystical philosophy of "pathetism," Sunderland might have taken rank with Braid

[1] Laroy Sunderland's "Book of Psychology."

and Liébeault as a pioneer of scientific psychotherapy.

But nothing did so much to discredit mesmerism among those in this country competent to investigate it as the fact that it soon became mixed up with spiritism. A certain clairvoyant ability had long been attributed to mesmerized subjects, and when, after Andrew Jackson Davis published his trance revelations from the " spirit world," and the Fox sisters began their spectacular career as " spirit rappers," clairvoyance became a leading feature of spiritistic séances, it was only natural that mesmerism and spiritism should be erroneously identified. To increase the confusion in both the popular and the scientific mind, many of the most prominent mesmerists joined the ranks of the spiritists, Laroy Sunderland in particular signalizing his conversion by the establishment, in Boston, of a spiritistic newspaper, *The Spiritual Philosopher*. This of itself was enough to condemn mesmerism in scientific opinion and to leave its practice entirely in the hands of traveling showmen and unscientific enthusiasts who used it more or less successfully in the treatment of disease.

It was by one of these obscure practitioners — Phineas Parkhurst Quimby — that the principles underlying religious psychotherapy were developed. Quimby was a clock-maker, a man of humble origin and scant education, but possessed of considerable native talent and force of character. He became interested in mesmerism through attending a demonstration given in his home town of Belfast, Maine, in 1838. So profound an impression did it make on him that he at once began to study it, and before long was able to mesmerize a good proportion of those who allowed him to experiment with them.

For a time he had no idea of turning his gift to therapeutic purposes. He simply used it, as so many other mesmerists did, to entertain and mystify. Nor did he theorize about it to any extent, beyond holding a vague opinion that it was some form of electrical action. But his interest deepened and his theorizing became more active when one of his best subjects, a young man named Lucius Burckmar, claimed to be able, when mesmerized, to look directly into the human body, see the organs at work, and treat any diseased conditions he found existing there. To the honest Quimby such a claim seemed preposterous, but he soon discovered that in a number of cases Burckmar actually succeeded in making a correct diagnosis and effecting a cure, usually by prescribing some simple remedy. Quimby himself had long been in poor health, and it occurred to him to test Burckmar's powers on his own account.

"He told me," he writes in a statement describing the startling result of his experiment, "that my kidneys were in a very bad state — that one was half consumed and a piece three inches long had separated from it, and was only connected by a slender thread. This was what I believed to be true, for it agreed with what the doctors told me, and with what I had suffered; for I had not been free from pain for years. My common sense told me that no medicine would ever cure this trouble, and therefore I must suffer till death relieved me. But I asked him if there was any remedy. He replied: 'Yes; I can put the piece on so it will grow, and you will get well.' At this I was completely astonished, and knew not what to think. He immediately placed his hands upon me, and said he united the pieces so they would grow. The next day he said they had grown together, and from that day I never have experienced the least pain from them." [1]

Then Quimby, as he expresses it, " began to think." He did not for a moment believe that the mesmerized Burckmar had really seen the diseased organ. He suspected, rather, that Burckmar had pictured it merely as the sufferer himself imagined it must look; and from this he leaped to the novel and startling conclusion that, so far as he

[1] Horatio W. Dresser's "Health and the Inner Life." This contains an excellent account of the life and teachings of Quimby, and of New Thought principles in general.

had had any disease at all, it was the result of his own thinking, and had been cured by nothing more than a change of thought. From this it was only a step to the assumption that diseased bodily conditions are invariably the effect of erroneous mental conditions, and may be overcome by getting the patient into the correct mental state.

So convinced was Quimby of the truth and importance of this view of disease that he determined to devote the rest of his life to promulgating it and to healing the sick by purely mental means. He dismissed Burckmar, and after several years of experimentation worked out an entirely new method of psychotherapy. Instead of mesmerizing a patient, he simply sat by his side and, after giving him a detailed description of his malady, impressed upon him the idea that the means of cure lay within himself, and that if he would only think himself healthy he would become healthy. The arguments he used to drive this home were, as the published extracts from his manuscripts show plainly, illogical and weak. But his earnestness went far to inspire conviction in the mind of a sufferer, and in numerous cases conviction was actually followed by cure.

It was Quimby's hope to develop his great " discovery " into a " science of health and happiness " that would bring comfort to all mankind. It would also seem that he contemplated putting his " science " on a religious basis, for he repeatedly declared that the " Truth," as he taught it, was identical with the teachings of Christ, and that Christ's miracles of healing illustrated and confirmed the principles which he advocated. But he did not live to diffuse the new gospel. Among his patients, however, were several willing and eager to carry on the work he had begun — if, perhaps, to continue and extend it along lines undreamed of by him. One of these patients, Mrs. Mary Eddy, became the founder of Christian Science. To two others the launching of the New Thought movement is due.

Mrs. Eddy had been cured by Quimby of

a malady of years' standing. Profoundly grateful, and readily acquiescing in his belief that he had made a discovery of the greatest significance to humanity, she joyfully accepted him as the prophet of a new dispensation, and with almost fanatical zeal set herself to study the " Truth " as this prophet had propounded it.

Little by little — but just at what time it is impossible to say, so shrouded in doubt and controversy is this phase of her career — she began to question the correctness of Quimby's explanation of the cures he worked. He was right, she felt, in teaching that disease was due to wrong thinking and could be overcome by getting the mind thinking right. But in her opinion it could be so overcome only because it actually was non-existent, the mind falsely imagining that the body was diseased. Thus, while Quimby had always conceded the reality of disease, although insisting on its mental origin, his disciple boldly affirmed its unreality. More than this, continuing her " investigations," she ultimately was led to deny the reality, not only of disease, but also of suffering, sin, and evil, and even of all things material; and took her stand squarely on as ultra-idealistic a philosophy as the mind of man has been invited to grapple with.

Its complete formulation was the work of years, and, we may well imagine, was attended by much brain-racking effort to meet the objections of worldly common sense. It is not necessary in the present connection to examine it in detail or to point out its many logical inconsistencies. What is important to note is the fact that Mrs. Eddy, after testing with some success her own powers as a healer, became convinced that any one sincerely and fully accepting her revised version of the Quimbyian gospel would thereby free himself from disease, and might confidently undertake the treatment of others; and she accordingly resolved to devote the remainder of her life to the propagation of her views. The result was the founding of a new religion.

Putting aside for a moment all considerations of its spiritual and therapeutic value — for Christian Science is essentially a religion of healing — it is impossible to resist a feeling of admiration for the courage, determination, and tireless energy displayed by Mrs. Eddy in her labors to gain a hearing. When she began her crusade she was a woman well advanced in years, of the scantiest means, and quite unknown. She had alienated many of her best friends by her devotion to her "queer ideas," and was practically alone in the world — a gaunt, sad, pathetic figure. Her first attempts at proselytizing only elicited derisive laughter. Yet she patiently persevered until at last, in the person of a young Massachusetts man, Richard Kennedy, of Amesbury, she found a convert willing both to adhere to the faith she preached and to aid her in making it known.

Together they opened in Lynn a school for the teaching of Christian Science, and, while Mrs. Eddy spent most of her time at work on her now world-famous book, " Science and Health," Kennedy sought to attract pupils by giving practical demonstrations of the therapeutic virtue of the doctrines he had learned from her. As a healer he proved successful enough to arouse a lively interest in the subject among the humble shoe-workers of Lynn, from whom his clientele was chiefly drawn, and it was not long before he had a number of applicants for instruction in " divine healing." This marked the turning of the tide, although it was not until several years later — after the publication of " Science and Health " and Mrs. Eddy's removal from Lynn to Boston, where she organized the First Church of Christ, Scientist, and established the Massachusetts Metaphysical College — that Christian Science took firm root and began to grow with the phenomenal rapidity that has won for it, within little more than a quarter of a century, a conspicuous place among the religious denominations of the United States.

In 1882, when Mrs. Eddy settled in

Boston, there were not one hundred Christian Scientists in the entire country. To-day there are almost one hundred thousand,[1] of whom four thousand are actively engaged in the work of healing. The movement has spread to foreign lands, and thus far shows no sign of diminishing vitality. On the contrary, every year sees numerous accessions to the ranks of those seeking salvation along the lines laid down in "Science and Health," and ardently subscribing to its uncompromising denial of the facts of the physical universe.

The same may be said of the New Thought movement, which has developed side by side with Christian Science. Its adherents also number far into the thousands, and it, too, has been growing increasingly influential. Unlike Christian Science, however, it has never become organized into a religious system, although it is distinctly religious in

---

[1] According to the latest religious statistics gathered by Dr. H. K. Carroll and published in *The Christian Advocate*, there were 668 Christian Science churches in the United States in 1909, with a total of 85,096 members.

character, and in some important respects its doctrines closely resemble those entertained by the followers of Mrs. Eddy. It upholds, as Christian Science does, an idealistic interpretation of life; it affirms the supremacy of mind over matter and the practicability of curing disease by purely mental means; and it finds warrant for its beliefs in the teachings of the Bible, particularly as exemplified in Christ's miracles of healing. But it parts company with Christian Science in refusing to acknowledge the validity of the latter's manifold " denials."

While the Christian Scientist denies the reality of the physical universe, the New Thought believer, to quote one of its best known exponents, Charles Brodie Patterson, " looks upon the visible universe as an expression of the power of God. He perceives that there must be an outer as well as an inner; that there must be effects as well as causes; that all the great material universe is the visible word of God — God's word becoming manifest in material form; that the body of man, to some degree, represents man's spiritual and mental life; that by the influx of man's spiritual consciousness the mind is renewed, and the body strengthened and made whole."[1] So, likewise, with disease, suffering, and sin, the reality of which is conceded by the New Thought, while claiming that they may be overcome by "the introduction of true thought into the mind of man." Consequently, the New Thought healer makes it his special business to introduce this "true thought" into the minds of his patients, confident that this is quite enough to cure them of disease.

Or, to put it otherwise, the New Thought harks directly back to Quimby's "get yourself thinking right." Indeed, it frankly acknowledges its indebtedness to Quimby, another point wherein it differs from Christian Science, which has long since repudiated him as an "ignorant mesmerist." The "father" of the New Thought movement, like the founder of Christian Science, was,

[1] Charles Brodie Patterson's "The Will to be Well."

as has been said, one of his patients, Warren Felt Evans by name, and formerly a Methodist clergyman. Less speculative than Mrs. Eddy, but sharing her belief that Quimby had fully demonstrated the possibility of healing disease "through the power of a living faith," Evans opened a "mind cure" sanitarium in western New Hampshire, and, besides treating those who came to him, wrote a number of books describing the benefits to be derived from practical application of the "spiritual laws" discovered by Quimby. "The Mental Cure," "Mental Medicine," and "Soul and Body" are the titles of the earliest of these books, all three of which, it is perhaps worth noting, were published before the appearance of Mrs. Eddy's "Science and Health." At the time, however, they attracted little attention, and the New Thought movement cannot be said to have fairly established itself until another patient of Quimby's — Julius A. Dresser, the father of Horatio W. Dresser, himself one of the most prominent New Thought writers of to-day — began to practice mental healing in Boston the same year that Mrs. Eddy removed to that city from Lynn.

Since then its growth has kept pace with, if it has not exceeded, that of Christian Science. Although handicapped to a certain extent by the absence of any formal organization such as that into which Christian Science has been welded, it has enjoyed the advantage of enlisting in its support a far larger number of able advocates than its great rival has ever secured; writers, for example, like Ralph Waldo Trine, Henry Wood, Aaron Martin Crane, and the younger Dresser, skilled in the art of presenting abstruse themes in language understandable by the average man. Moreover, its explicit recognition of the material side of life has commended it in quarters where the sweeping negations of Christian Science arouse only a feeling of contempt. But the principal reason for its success is found in the fact that, notwithstanding its doctrinal crudities and extravagances, it has proved sufficiently "workable" to justify, in the opinion of its adherents, the extreme claims it puts forth.

The same must be conceded of Christian Science. While it is lamentably true that the Christian Science healer has been guilty of much serious malpractice, it is equally certain that he has effected cures in cases pronounced hopeless by orthodox practitioners. And it is incontestable that in numerous instances Christian Science believers, as also followers of the New Thought, appear to gain greatly in health and happiness, growing more robust, efficient, energetic, and contented than they were before their "conversion." All this, of course, is most helpful in the way of winning recruits, and goes far to wring even from the obdurately skeptical a reluctant admission that "there may be something in it, after all."

In order to appreciate just what that "something" is, and to understand why Christian Science and the New Thought, on their therapeutic side, are so strangely compounded of success and failure, it is necessary to take account of the progress achieved by an altogether different type of mental healers — men of scientific temperament and training, whose efforts have been directed to upbuild a system of psychotherapy based, not on mystical speculation, but on exact knowledge. In fact, were it not for them, psychotherapy, so far as concerns any real comprehension of its workings, would still be pretty much where it was in the dark ages of Mesmer. While others have been rashly conjecturing, they have quietly investigated, experimented, and observed; and although they are few in number, and have been at work only a comparatively short time, they have already made remarkable headway in fathoming the processes of mental healing, and in determining its proper place in the practice of medicine.

II

Principles and Methods

THE foundations of scientific psychotherapy may be said to have been definitely laid about thirty years ago, when men of good repute as physicians and psychologists began for the first time to make an organized investigation of the phenomena of hypnotism, scientific interest in which, as already stated, had been aroused in France by Liébeault's demonstration of its therapeutic helpfulness. Two great centers of experimentation were established, one in the town of Nancy, under the supervision of Liébeault himself, the other in Paris, at the asylum of the Salpêtrière, then in charge of the famous Dr. Charcot.

At both these places it was soon ascertained that, quite apart from its power as a healing agent, suggestion, when applied in the hypnotic trance, was capable of producing most extraordinary effects on the human organism. It could seriously modify the processes of nutrition, circulation, and digestion; could bring about temporary loss of the power of sight, speech, hearing, feeling, and motion; and could even cause the appearance of blisters, swellings, eruptions, etc., on the body of the entranced subject. The mental apparatus was affected most remarkably. Under hypnosis patients were able to remember incidents in their past life which had vanished completely from their waking consciousness; and, more striking still, if, while hypnotized, they were given suggestions that involved the performance of some act at a specified time in the future, they would faithfully obey these " post-hypnotic " commands, notwithstanding the fact that when dehypnotized they knew nothing of the suggestions they had received.[1]

---

[1] The phenomena of hypnotism will be examined in some detail in the section, "Hypnotism as a Therapeutic Resource."

It seemed a legitimate inference that there existed a much closer relationship between the psychical and the physical in man than had previously been suspected, and that, in view of the effects of hypnotic suggestion on the physiological processes, it was possible that many maladies apparently physical in character had their origin in some psychical disturbance and could best be treated by psychical means. Verification of this theory was not long in forthcoming. Among the patients at the Saltpêtrière were a number of victims of hysteria, a disease which, on account of the predominance of such symptoms as convulsions, paralyses, and contractures, had been regarded as primarily physical rather than mental, and treated accordingly, with but little success. By hypnotizing these patients and calling up in hypnosis the memories of their past life, Charcot and his fellow psychopathologists [1] were able to locate the source of all their troubles in long-for-

[1] This is the term now generally applied to those students of abnormal psychology who conduct their researches with therapeutic ends in view.

gotten experiences — frights, griefs, and so forth — which in some subtle way had thrown the nervous system out of gear and provoked the hysterical attacks.

Having thus demonstrated the distinctly psychical nature of one disease — and having incidentally learned the value of hypnotism for diagnostic as well as therapeutic purposes — the investigators broadened their field of inquiry, and gradually discovered that besides hysteria there were numerous maladies which similarly originated from psychical disturbances of one kind or another. The disquieting experience might have passed completely from the recollection of the sufferer, yet under hypnosis it readily revealed itself as existing subconsciously in his memory and acting as a perpetual irritant to produce all manner of unpleasant symptoms, physical and mental. In all such cases it was found that a cure could be effected by suggestion when ordinary methods of therapy were of little or no avail.

But, what is most important, it was also ascertained that the efficacy of suggestion itself often depended on the precision with which a diagnosis was made and the secret, psychical cause of distress brought to light. Nor would suggestion succeed if the "dissociation," as it is called, had progressed so far as to involve radical destructive changes in the nerve cells, rendering the malady "organic" instead of merely "functional." For, as the psychopathologist frankly admits, suggestion is powerless in the presence of all "organic" diseases, whatever their origin, or is at best useful as an auxiliary to their treatment by physiological, chemical, and surgical methods.

On the other hand, he has learned that not infrequently the dissociative process gives rise to symptoms simulating those of organic diseases, particularly in the case of sufferers from hysteria. Some hysterical affections, for example, are easily mistaken for tuberculosis of the lungs or other organs, for abdominal and uterine growths, for intestinal obstructions; and if the patient happens to be attended by a physician unacquainted with the myriad forms in which hysteria may show itself, a wrong diagnosis is certain to be made, often with tragic consequences that would have been averted had the true character of the disorder been recognized and resort had to psychotherapy. As Dr. Pierre Janet, one of the foremost of living psychopathologists, pointed out in a course of lectures delivered at the Harvard Medical School, it is impossible to estimate the number of unnecessary and useless operations that have been performed to remedy conditions which really called for treatment by suggestive therapeutics.

Still further, at an early stage of their experiments the Nancy investigators discovered that in some cases suggestion might be utilized therapeutically without the aid of hypnotism. This in turn led to the discovery that every one is more or less suggestible, and rendered possible the development of a system of non-hypnotic psychotherapy resting on scientific knowledge of the laws of suggestion as worked out by painstaking psychological analysis.

To-day, consequently, the scientific psychotherapist does not feel obliged to make such extensive use of hypnotism as in the days of Liébeault and Charcot, but frequently works directly on the waking consciousness of his patients, deftly applying therapeutic suggestions by methods that vary according to the particular requirements of the case. There even are some psychotherapists of the scientific type — such as Dubois, of Berne — who seem to find it unnecessary to use hypnotism at all. The majority, however, employ it to a greater or less extent, especially for purposes of diagnosis, it being their experience that only through hypnosis — or kindred methods to which reference will later be made — is it possible to get at the subconscious mental states in which so often lies hidden the real cause of the malady they are endeavoring to cure. And, whether they utilize hypnotic or non-hypnotic suggestion, all scientific psychotherapists are agreed in recognizing that suggestion has its limitations, and that within those limitations it is necessary for the suggestionist to be thoroughly grounded in the psychological principles governing its action in order to be able to apply it with any certainty of success.

Herein is the great difference between scientific psychotherapy and the psychotherapy practiced by Christian Science and New Thought healers. Where the latter succeed they owe their success, equally with the scientific psychotherapists, to the influence of suggestion. Where they fail it is because they ignorantly treat diseases not susceptible of cure by suggestion; or because, in cases where a cure may be thus wrought, they lack the training that would qualify them to make a precise diagnosis, ascertain the true cause of trouble, and overcome it by one or another of the various methods at the command of the scientific practitioner. Fortunately for them — and for their patients — suggestion, even when unguided by scientific knowledge, is often sufficient of itself to work seemingly miraculous cures. In such cases all that is needed is to imbue the sufferer with a profound conviction, a " lively faith," in the possibility of his regaining health.

This faith Christian Science and the New Thought inspire by their appeal to the religious side of man's nature, by emphasizing the goodness of God, and by systematically cultivating in their adherents a spirit of hopefulness, buoyancy, and courage. So long as they can do this it matters not, from the therapeutic point of view, how erroneous their doctrines may be. Right or wrong, the result is the same — the suggestibility of the believer is increased to a point which renders him peculiarly responsive to curative ideas, and, if he is suffering merely from some functional complaint, may bring about his complete recovery. There is always, however, the danger that his trouble may be organic instead of functional, in which event, his last state is sure to be worse than his first.

But the far-reaching differences between the methods of the scientific and the non-scientific mental healer, may best be made clear by citing a few illustrative cases from the experiences of leading psychopathologists.

There was brought to the office of Dr. Boris Sidis, of Boston, a young man suffering from what were supposed to be attacks of that dread disease, epilepsy. He was a typical product of the slums, gaunt, hungry-looking, undersized. Born of parents of the lowest social strata, he had been treated from infancy with harshness and brutality. He had had no schooling, and could neither read nor write. Except for the names of the President and a few ward politicians, he knew nothing of the history of his country. All his life he had known only poverty and hard work.

And now it seemed that even the chance of earning a meager living by hard work was about to be taken away from him.

Principles and Methods

"I have such fearful shaking spells," he told the doctor. "They come on me day and night. I shake all over, my teeth chatter, I feel cold. Then I fall to the floor and lose my senses. Sometimes my fits last three hours."

"Have you had them long?"

"Yes, almost since my boyhood. But they are getting worse all the time."

After a careful examination and the application of the most rigid tests had revealed no sign of organic trouble, Dr. Sidis suspected that the convulsive attacks might be nothing more than the outward, physical manifestation of some deep-seated psychical disturbance. He questioned the young man closely:

"Can you remember just when these attacks began?"

"No."

"Did you have them when you were a child?"

"I don't think so."

"Was there anything that occurred during your childhood likely to leave a particularly disagreeable impression on you?"

"Why," he replied, "I have been unhappy all my life. As a boy I was beaten and kicked and cursed. But I don't think of anything special."

"Will you let me hypnotize you?"

"You can do anything you like to me, doctor, so long as it will help me get well."

But it was found impossible to hypnotize him — he was in too agitated, too excited a state.

Now, psychopathologists long ago discovered that not everybody was hypnotizable; and, moreover, that many persons would not permit themselves to be hypnotized. So they have been obliged to devise other means of "tapping the subconscious," Among these is a method known as hypnoidization, which results in putting the patient into a half-dozing, half-wakeful condition, wherein long-forgotten memories crop up in the mind.

Principles and Methods 51

Making use of this method, Dr. Sidis soon had his patient in a quiescent state — in fact, to all appearances asleep.

"Now," said he, in a low tone, "tell me what you are thinking about."

At first there was no response, but presently the young man began to talk. It was evident that he was recalling memories of his childhood — sordid, pathetic, almost tragic scenes.

He spoke of a "dark, damp cellar" in which, when a very little boy, he had been forced to sleep, and where it was bitterly cold. He spoke of the terror it had inspired in him, and how he had been afraid to go to sleep, lest he should be gnawed by rats. Then, with startling suddenness, he leaped out of his chair, shaking in every limb, teeth chattering, speech paralyzed. He was in the throes of one of his attacks.

The doctor nodded his head understandingly.

It was not an epileptic case. It was a typical instance of a seemingly purely physical malady having its origin in a psychic shock.

Consciously the sufferer had forgotten all about the nights passed in the cellar so many years before. They had utterly vanished from his waking memory. But subconsciously he remembered them as distinctly as though they were not past but present experiences — subconsciously he was continually living them over again, to the gradual breaking down of his nervous system, of which the convulsive attacks were symptomatic. In fact, it was found that they could be brought on merely by uttering in his hearing the words "dark" and "damp," which seemed to act as psychic triggers exploding the mine of horror memories in the depths of his subconscious being.

A few weeks of suggestive treatment directed to the complete blotting out of the disease-producing memories, and he was permanently freed from his terrible affliction.

More frequently, the symptoms in dissociational cases are wholly mental. Here is

Principles and Methods

a characteristic example, likewise taken from the experience of Dr. Sidis, who, it may incidentally be said, shares with Dr. Morton Prince, also of Boston, unquestioned preeminence among the few psychopathologists whom America has as yet produced.

A middle-aged gentleman resident in a New England town, highly educated, successful in business, and generally regarded as a man of great intellectual keenness and strength of will, called at his office one day and announced:

"Doctor, I have come to see you about a matter which may seem absurd, but which is making life a perfect hell to me. Put briefly, the trouble is that I am afraid to go out nights."

"By that you mean — ?"

"I mean that as soon as darkness sets in, I become a coward. I dare not stir from the house. No matter what imperative demands my business may make, no matter what social engagements I should keep, I simply do not dare to go outdoors.

"I do not know what it is that I am afraid of. It is just a vague, haunting, overpowering dread that seizes me as soon as night comes. My relatives have argued with me, I have argued with myself. I know it is absurd, but I simply cannot shake it off. And, doctor, I tell you it is killing me."

Putting him in the hypnoidal state, Dr. Sidis, note-book in hand, jotted down every word that fell from his lips.

Mere fragments of ideas they were, like the swiftly changing fancies of a dreamer. All at once he muttered:

"They will kill me! What a blow that was! I can never get home."

The psychopathologist bent forward, listening eagerly.

"How dark it is! How my head hurts! Yes, they got all my money."

And now, piecemeal but in graphic detail, he rehearsed an experience of his youth — an attack made upon him one night by two highwaymen, who had beaten him into unconsciousness.

Principles and Methods

In that attack lay the clue to his seemingly irrational fear.

He had apparently recovered from its effects, no physical harm had resulted. He had long since dismissed it from his mind. Yet subconsciously he had never forgotten it; subconsciously he was haunted by the idea that if he went out at night he would again be attacked by footpads!

He was like a man tormented by a perpetual nightmare, and, like the victim of a nightmare, he awoke to a full realization of the folly of his terror and was able to overcome it as soon as it was presented in its true light to his waking consciousness.

Precisely the reverse was the case of a woman who feared to leave her house not at night but in the daytime. In the normal, waking state she could give no explanation for this obsessing fear, but put into the hypnoidal state its explanation was soon forthcoming.

Years before there had come into her life one of those domestic tragedies of all too common occurrence. She had discovered that her husband was unfaithful to her, and that he had become infatuated with another woman.

Like many another wife she had kept her sorrow to herself. But the shock had so unnerved her that she began to imagine that everybody she met in the street knew of her troubles and was talking about them. Soon she could not bear to go outdoors, and became almost a recluse, appearing in public as little as possible.

After a time, however, there had been a reconciliation, and she became, to all outward seeming, happy and light-hearted as before, going everywhere, entering freely into social amusements, and apparently being in perfect health. Nevertheless, the bitter experience through which she had passed had left a deep psychic wound that never completely healed.

Without realizing it, she was constantly tormented subconsciously by the old idea that everybody she met was talking about her. From this, years afterwards, developed the seemingly inexplicable fear of going outdoors in the daytime.

Asked, while in the hypnoidal state, why she was not afraid to go out after dark, she promptly replied:

"Because in the dark no one can recognize me."

Subconsciously, in other words, the sorrow and the dread and the bitter thoughts of the period of alienation from her husband were still present experiences to her — were still as real and painful as at the time of their actual occurrence.

All this was revealed through hypnoidization, and a complete cure speedily effected, the baneful memory-images being rooted out of her subconsciousness, or, to speak more accurately, being "reassociated" with her upper consciousness.

Sometimes dissociational disorders result not from a single emotional disturbance but from a succession of psychic shocks, giving rise to the most complicated symptoms. I have in mind a recent striking case of this sort, in which, after years of indescribable suffering, a woman of sixty was by psychopathological treatment cured of lung, stomach, and kidney trouble, to say nothing of an extreme nervousness and an insistent fear that she was becoming insane.

When she applied for treatment she presented a pathetic appearance. She was haggard, emaciated, and weak, her skin dry and crackling, her heart action irregular. She had a racking cough, and occasionally, she said, suffered from convulsive attacks during which she became unconscious. But most of all she complained of sensitiveness of the stomach, of kidney trouble, and of nervousness.

"When the nervous spells are on me," she declared, "I suffer death agonies. I cannot sleep, I cannot eat, my head feels as though it would burst. Time and again I have been on the verge of committing suicide.

"Then, too, I feel as though I must be going crazy. Though I can read and study and take up any intellectual pursuit without the slightest ill effect, if I attempt, for instance, to buy a dress for myself, my brain gets on fire and I walk the floor in a frenzy of excitement, quite unable to decide what choice I should make. Yet I experience no difficulty in making purchases for other people, and my judgment is considered so good that my friends often ask me to help them in their shopping. And I cough, day and night, sometimes for hours together."

A thorough examination, however, failed to disclose any indication of organic lung disease, or of kidney or stomach disease. Besides which, unlike the young man with the "epileptic" seizures, the patient was found to have an excellent family history, from the medical point of view. Both her father and her mother had been of rugged constitution and had lived to a good old age. Dissociation was at once suspected, and she was hypnoidized.

Almost the first statement she made in the hypnoidal state related to a long-forgotten incident of childhood that had been the starting-point of all her troubles.

At the age of five — fifty-five years before she sought psychopathological aid — she had been frightened into an hysterical attack by the sight of an insane woman in a maniacal state. For months afterwards the image of that woman never left her mind, and she kept asking herself, "Do little girls go insane?"

And even after the image faded from her waking memory it remained as vividly as ever in her subconsciousness — as was shown by the fact that, although before being hypnoidized she had stated that she never dreamed, in the hypnoidal state she remembered that she frequently dreamed an insane woman was standing near her bed, bending over her.

To this subconscious memory-image, persisting all unknown to her for more than half a century, was due her unconquerable fear that she would herself some day become insane.

Another horror memory that had affected her whole after-life was connected with an occurrence of her early girlhood. At the age of eleven she had been frightened into insensibility by the action of a girl friend in dressing up as a "ghost" and darting out upon her in a dark room. In her waking state she remembered nothing of this; hypnoidized, she recalled it vividly.

When eighteen, having become a school teacher, she had worried greatly because of failure to secure promotion. From this period dated her headaches, as well as her first serious nervous attack.

But the culminating shock — the experience to which her physical ills were chiefly due — was sustained in middle life, when her only daughter, after growing up to womanhood, fell a victim to consumption. Throughout the weary weeks of her daughter's illness she watched in anguish at her bedside. The distressing cough, the gastric disturbances, the loss of appetite, the nausea, the inability to retain food — every symptom seared itself into the mother's subconsciousness, never to be forgotten and eventually to be reproduced, by the strange power of subconscious mental action, in the mother herself.

Caused by the mind they were curable by the mind. One by one the psychopathologist attacked and eradicated these deadly subconscious memories, and with their blotting out the patient's health constantly improved, until at last the entire complex of symptoms had disappeared.

Here, then, we find subconscious mental action responsible for the production of seeming insanities, delusions, irrational fears, and, in the case of this unhappy woman of sixty, even causing the appearance of symptoms resembling those of true organic disease.

Finally, to mention a typical instance in which a wholly unnecessary and useless operation was averted by psychopathology, there is the case of a young woman of Providence, R. I., whom a lucky chance took to a

Principles and Methods 63

neurologist, Dr. John E. Donley, an ardent student of psychopathological methods.

She had been sent to him by her physician to determine what particular nerve in her hand ought to be "resected" to relieve a semi-paralysis from which she had been suffering for some time. A year or so before she had been bitten in the hand by a pet cat. At first she had felt no ill consequences, the wound healing nicely. But after a time a pain had set in, gradually extending up the arm, which had become almost helpless. It was her physician's opinion that some nerve had been caught in the scar of the wound, and that an operation, which she greatly dreaded, would be necessary to restore the arm to usefulness.

Before examining her hand Dr. Donley decided to make a psychopathological examination as to her general nervous condition. The discovery immediately followed that the paralysis of her arm was nothing more than an hysterical disturbance.

Hypnoidizing her, he found that the attack made on her by the cat had caused a profound psychic shock. She had been almost panic-stricken with fear, insisting that blood poisoning would surely result; and, although the wound had healed as her physician predicted it would, she still subconsciously clung to this idea.

What she required was not the surgeon's knife but treatment by suggestion. Only a few such treatments were needed to work a complete cure.

But — and this is a point that cannot be emphasized too strongly — even suggestion would in all probability have failed had not the neurologist been able, by the methods of psychopathological diagnosis, to get at the exact cause of the trouble and apply precisely the suggestions needed to meet the situation.

This it is that most sharply differentiates scientific psychotherapy from the psychotherapy of the faith healer. To repeat what was said above:

Both the scientific psychotherapist and

the faith healer make use of suggestion to attain their ends. Both get results, for the reason that suggestion, even when utilized by an untrained practitioner, is frequently powerful enough to bring about seemingly miraculous restorations to health.

But whereas the non-scientific psychotherapist, with few exceptions, applies suggestion indiscriminately to all manner of diseases, the scientific psychotherapist recognizes that it is by no means a cure-all, and that even in cases where it is beneficial a thorough, accurate diagnosis is often indispensable to a perfect cure.

As between these two types of psychotherapy can there be any doubt which is the " true mental healing " — that which takes its stand on blind faith, or that which depends on the proven facts of scientific experiment and observation?

III

Masters of the Mind

I HAVE already given an outline account of the wonderful new science of psychopathology, or medical psychology, and of the development by its aid of a scientific system of mental healing which the physicians of this country, as of other lands, are beginning to adopt. Now I want to say something about the men who, by their investigations and remarkable cures, have done most to convince the medical world that the human mind possesses powers which, when scientifically directed, are almost incredibly efficacious in conquering many widespread and hitherto baffling diseases.

They are an exceedingly interesting group, these premier psychopathologists. There are four of them, representing by birth as many countries — France, Austria, Russia, and the United States. But the Russian in early manhood made his way to this country, so that, of the four leaders of scientific mental healing, two are Europeans and two Americans. Their names are Pierre Janet, of Paris; Sigmund Freud, of Vienna; Morton Prince, of Boston; and Boris Sidis, of Boston.

Of the four, I must speak first of Janet. He it was, who, under the inspiring guidance of the famous Dr. Charcot, first called attention to the importance of psychology as an aid in the practice of medicine, and made the marvelous discovery of the rôle played by mental experiences of an emotional nature in the causation of many diseases. Yet, curiously enough, he began his professional career without any idea of becoming either a psychologist or a physician.

His great ambition, cherished from early youth, was to win a name as a philosopher. Graduating from a Parisian college with high honors, he was, in 1881, when only twenty-two years old, appointed Professor of Philosophy in the College of Chateauroux, and afterwards received a similar appointment in the College of Havre. But in the meantime he had become interested in the experimental investigations of hypnotism begun in the town of Nancy by Drs. Liébeault, Bernheim, Beaunis, and Liégeois, and by Dr. Charcot at the Salpêtrière, that great refuge for the sick and destitute of Paris. In hypnotism Janet thought he saw an unrivalled instrument for studying the nature of men; and, returning to Paris, he entered the Salpêtrière as a pupil of Charcot's — a pupil who was soon to excel his master.

He found that Charcot had brought together, for clinical study, what a visitor to the Salpêtrière once described as "the greatest collection of hystericals the world has ever seen." Up to that time it had been generally believed that hysteria was a physical malady associated with, and resulting from, some organic trouble. Charcot's investigations had proved that this was entirely wrong. Still more important, Charcot had vastly broadened the medical conception of hysteria by showing that quite frequently maladies diagnosed as organic and incurable were in reality nothing but hysterical affections.

Thus, patients were brought to him who had not uttered a word for years, but when hypnotized spoke fluently; while others, supposed to be paralytics, walked with ease during the hypnotic trance, and sometimes during natural sleep. In one very striking case a patient, who had long been suffering from a paralysis of the legs, got out of bed one night in a somnambulic state, seized his pillow, which he held tightly pressed to his breast as though it were a child, fled into the hospital courtyard, and climbed nimbly up a gutter-pipe and up the sloping roof of one of the buildings. An attendant who ran after him was quite unable to climb the roof, and had much trouble in persuading him to descend; and when, after having come down from his dizzy perch, the attendant awoke him, he instantly became paralyzed as before, and had to be carried back to bed!

But Charcot did not live to round out his epoch-marking labors by discovering the mechanism of hysterical affections and their proper treatment. This it remained for Janet to do. What he saw in the Salpêtrière so inspired him with a desire to help the human wrecks that thronged its wards, that he abandoned all thought of metaphysical achievements, and resolved to enlist in the battle against disease. It is not my intention to describe in detail the investigations which ultimately convinced him that hysteria was the product of emotional experiences, and that it could be cured by mental means; but I would give a few instances that will bring the facts out with sufficient clearness, and satisfy the reader as to the vital importance of this discovery.

A girl of eighteen once applied at the Salpêtrière for treatment for convulsive attacks from which she had been suffering for two years. They came on at irregular but increasingly frequent intervals, and invariably began with a fainting-fit. As consciousness gradually returned she would utter piercing shrieks of terror, with cries of "Lucien! Lucien!"— as if appealing to some one to defend her. Then she would rush to the nearest window, throw it open, and lean out, calling "Thieves! Thieves!" After this she would immediately reënter her normal condition, knowing nothing of what had occurred during the convulsive attack.

Dr. Janet suspected that the scene which she thus dramatically enacted was reminiscent of some disastrous experience of her earlier life, and was the direct cause of her hysteria; but the girl assured him that she knew nobody named Lucien, and could not recall anything that had ever given her such terror as she displayed.

Put into the hypnotic trance, however, the patient remembered that some years before she had been offered a grievous insult from which a certain Lucien had defended her; and that, a few days afterwards, thieves had broken into the château where she was working. The emotional shocks caused by these experiences were responsible for the convulsive, somnambulic attacks; which, in turn, had obliterated all recollection of the original experiences from the girl's waking memory. Still more remarkable, the convulsive attacks ceased the moment Janet succeeded in making her remember the episodes that had caused them.

In another case that gave him far more trouble, the patient suffered from a persistent hallucination of seeing a man in the room with her. Her relatives believed that she was insane, and wished to place her in an asylum, as she occasionally manifested suicidal tendencies. But Dr. Janet diagnosed her case as one of hysteria, and with the aid of hypnotism made the interesting discovery that the hallucinatory image which she thought she saw was the figure of a lover who had deserted her several years before. It appeared that every time she thought of her faithless sweetheart, his image rose before her.

To Janet it seemed a perfectly simple matter to "suggest" away the hallucination, by impressing upon her, during hypnosis, the idea that when she awoke she would no longer see the imaginary form. But he found that for some reason the suggestion would not "take." Day after day he patiently hypnotized her, always without success. Finally, he began to suspect that at bottom she did not want to be cured, and that the passionate desire to see her lover if only as a phantasm constituted too strong a "self-suggestion" to be overcome by direct attack. Another method would have to be tried.

"Very well," he one day said to her, while she was hypnotized, "if you want to continue seeing your lover, you shall see him. But, remember, you will always see him with the head and face of a pig."

He then brought her out of the hypnotic sleep into her natural state. Five minutes later she uttered a cry, and covered her eyes with her hands.

"What is the matter?" inquired Janet, calmly.

"It is terrible! Terrible!" she exclaimed. "I see a man standing in the corner of the room, and his face is like a pig's!"

"How absurd!" said Janet.

After this, he left her to her own devices, no longer hypnotizing her. For a few days she complained that everywhere she went she saw the man with the face of a pig. Gradually the hallucinatory image faded, and at length entirely disappeared, leaving her restored to perfect health. As Dr. Janet afterwards explained, the grotesque hallucination which he had succeeded in impressing upon her, had brought about a profound revulsion of feeling. Manifestly, she could not love a man with a pig's head. She no longer wanted to see her sweetheart, or to think of him, and in proportion as she ceased to think of him, the hallucination disappeared.

This method Janet calls the method of substitution, but it is only one of several methods used by him. Like every good physician he varies his methods to suit the requirements of the case.

The point on which he insists, is that in dealing with hysteria and other maladies curable by mental means the great thing is to recognize that they are invariably conditioned by mental states;[1] and that, in order to be sure of working a cure, it is necessary to get at the underlying subconscious ideas and eradicate them. Furthermore, he lays stress on the tremendously important fact that profoundly distressing emotional experiences of the kind just indicated do not always give rise only to mental and nervous symptoms, but frequently cause most

[1] Dr. Janet's views are clearly set forth in his book "The Major Symptoms of Hysteria," which contains his Harvard Medical School lectures on the subject. See also his earlier works, especially the "État Mental des Hystériques," "Névroses et Idées Fixes," and "Les Obsessions et la Psychasthénie."

appalling physical disorders, curable, however, by the methods of psychopathology.

In the case of the paralyzed roof-climber, for instance, Janet learned that the paralytic, who was a widower, had had violent quarrels with his mother-in-law over her treatment of his only child. It was after one of these quarrels that his paralysis had set in, as the result of a panicky, irrational, subconscious fear as to what would happen to the child if he should ever be unable to rescue it from the clutches of its wicked grandmother. In this way he had unwittingly suggested to himself the idea of paralysis, and, since he was of an unstable, neurotic temperament, the suggestion had proved so powerful that he had actually become paralyzed. The roof-climbing incident at the Salpêtrière, like the scene enacted by the girl with the convulsive attacks, was reminiscent of the cause of the paralysis, and pointed the way to its successful treatment.

In appearance Janet is a rotund, robust, merry-faced little Frenchman, with a rich fund of humor, sensible, and practical. There is nothing in him of the visionary or the fanatic. So likewise with our two American psychopathologists, Drs. Prince and Sidis.

The latter, as his name implies, is of Russian birth, but all his scientific work has been done in the United States, to which he came, while still a very young man, after some thrilling experiences in his native land, where he had become involved in the revolutionary movement; had been arrested, clapped into a fortress, and narrowly escaped a sentence to Siberia. Following his release the police made matters so uncomfortable for him that he fled the country, and, after a brief sojourn in Germany, made his way to New York in 1888, knowing not a word of English, friendless, and almost penniless.

Less than a decade later — the young Russian having managed to put himself through Harvard, where he came under the stimulating influence of Prof. William

James, and was led to specialize in psychology — he astonished the veterans in that science by the publication of a striking book on "The Psychology of Suggestion." In the meantime he had been appointed Associate in Psychopathology in the then recently established Pathological Institute of the New York State Hospitals. Here he remained several years, developing his method of hypnoidization and effecting many impressive cures.

One of these may well be given to illustrate with increased emphasis the subtle and far-reaching influence of the mind in causing disease, and the diagnostic and therapeutic value of hypnoidization.

There was brought to Dr. Sidis, as a last resort before committing the sufferer to an asylum, a young man of twenty-five who presented as complex and astonishing a combination of symptoms as is to be found in medical annals.

He was afflicted, for one thing, with an insistent belief that he was always making mistakes, even with regard to the most trifling matters. If, for instance, he wrote a letter, he was never sure that he had addressed it correctly, and others had to read the address over in order to satisfy him. In locking his bedroom door, he had to try the lock over and over again, to get full assurance that he had really locked it. When retiring he never felt certain that he had turned off the gas-jet, and felt obliged to get up and test it with a lighted match. Besides this perpetual "*folie de doute,*" he was troubled with an absurd desire to "tear out his eyes, put them under a weight, and have them crushed." He frequently suffered, too, from brief attacks of psychic paralysis, or "aboulia," feeling temporarily deprived of all power of speech and motion.

Nor does this exhaust the catalogue of his ills. He had an irrational fear of contracting some deadly disease, more particularly consumption, and was forever washing his hands "to rub the germs off." He complained of a palpitation of the heart, and was unquestionably troubled by a chronic irritation of the bladder, which caused him a great deal of inconvenience, and which ordinary medical treatment had utterly failed to relieve. Altogether, his condition seemed to be hopeless, and such as to justify the fear of his family that he was doomed to spend the remainder of his life behind the walls of an institution.

But Dr. Sidis, by the application of some delicate tests, ascertained that, whatever the nature of his complicated malady, the unfortunate young man was not really insane. The likelihood, therefore, was that his entire complex of symptoms, physical as well as mental, was actually nothing more than the outward manifestation of unpleasant subconscious ideas, associated with forgotten experiences of his earlier life. To get at these subconscious ideas, Dr. Sidis made use of his method of hypnoidization.

Here is his own account of the manner in which he puts his patients into the hypnoidal state:

"The patient is asked to close his eyes and keep as quiet as possible, without, however, making any special effort to put himself in such a state. He is then asked to attend to some stimulus such as reading or singing (or to the monotonous beats of a metronome). When the reading is over, the patient, with his eyes shut, is asked to repeat it and tell what comes into his mind during the reading, or during the repetition, or immediately after it. Sometimes the patient is simply asked to tell the nature of ideas and images that have entered his mind. This should be carried out in a very quiet place, and the room, if possible, should be darkened so as not to disturb the patient and bring him out of the state in which he has been put.

"As modifications of the same method, the patient is asked to fix his attention on some object, while at the same time listening to the beats of a metronome; the patient's eyes are then closed. After some time, when his respiration and pulse are found somewhat lowered, and he declares that he thinks of nothing in particular, he is asked to concentrate his attention on a subject closely relating to the symptoms of the malady.

"The patient, again, is instructed to keep very quiet, and is then required to look steadily into a glass of water on a white background, with a light shining through the contents of the glass; a mechanism producing monotonous sounds is set going, and after a time, when the patient is observed to have become unusually quiet, he is asked to tell what he thinks in regard to a subject relating to his symptoms. In short, the method of hypnoidization is not necessarily fixed, it admits of many modifications; it is highly pliable and can be adjusted to the type of case as well as adapted to the idiosyncrasies of the patient's individuality."[1]

Simple as this process sounds, it has, as

---

[1] Boris Sidis's "Studies in Psychopathology" in the *Boston Medical and Surgical Journal*, vol. clvi.

was stated on a previous page, a peculiar effect, sending the patient into a half-waking, half-sleeping state — the hypnoidal state — during which he can recall, sometimes with startling vividness, memories of events and experiences which have long faded from his consciousness. It was thus with the young man whose case has just been outlined.

Fragmentarily but vividly a host of grim memory pictures floated into his mind, and were described by him as he lay hypnoidized. When he was a very young child, it appeared from the statements he made during hypnoidization, he had lived with an aged grandfather who had been a sufferer from a peculiarly distressing bladder trouble, had been remarkably absent-minded, and had had difficulty and hesitancy in handling anything given to him. All this the child had watched with great sympathy and grief. After his grandfather's death, however, he had gradually forgotten, so far as his conscious memory was concerned, all about the old gentleman and his troubles; but the impression made on his sensitive, imaginative nature had been too profound to allow the sad experiences he had witnessed to fade away completely. In other words, the young man's bladder trouble, his aboulia, and his *folie de doute,* were symptomatic of no organic malady but were purely functional, and were the " working out " of the painful emotional experiences of childhood, which subconsciously he had never forgotten, and which had been able to spring into baneful activity and develop into disease-symptoms as soon as he had weakened himself by overstudy.

So with his other symptoms. By means of the method of hypnoidization, his irrational fear of contracting consumption was traced back to his having witnessed, at a tender age, the death agonies of an aunt who had died of tuberculosis. His absurd desire to tear out his eyes and crush them had its origin in another experience of childhood, when he had an inflammation of the eyes and had to undergo the ordeal of having them bathed with various washes. During hypnoidization he also recollected having heard, when a child, horrible stories about people whose eyes "swell and bulge and then crack and break." One can readily imagine, as Dr. Sidis says, "what a deep and lasting though subconscious influence such gruesome tales may exert on the sensitive mind of a highly imaginative child."

Not all of these forgotten memories were recovered by a single hypnoidization. It required weeks of patient endeavor to bring them fully above the threshold of consciousness. But eventually Dr. Sidis had in his possession, so to speak, a complete map of the starting-points of his patient's symptoms, and was able to work an absolute and permanent cure.

All that he had to do, having once got at the specific disease-producing memories, was to recall them one by one to the young man's waking consciousness, showing them to him in their true light as mere memory-images of past events, and at the same time impressing upon him, through suggestions given in hypnoidization, the belief that they would henceforth have no ill effect on him.

Now, while he has been curing his patients, Dr. Sidis has also been studying them, and has reached some novel and startling conclusions. Chief among these is his doctrine of reserve energy.

According to this doctrine, each of us possesses a stored-up fund of energy, of which we ordinarily do not make any use, but which we could be trained to use habitually to our great advantage. Dr. Sidis contends that it is by arousing this potential energy that the patients whom he treats are cured; and he further insists that it is actually possible to train people to draw readily and helpfully on their hidden energies.

If he is right in this contention his psychopathological researches obviously have a vital bearing, not only on the problems of medicine, but on equally important problems in the domain of educational and social reform. In any event, it is conceded that by his masterly analysis of the laws of suggestion, his development of the hypnoidal state, and his classification of the factors governing the production of mentally caused diseases, he has made highly original and valuable contributions to the growth of the new science which seems to promise so much for the future of humanity.

And now to pass from Dr. Sidis to Dr. Morton Prince, who is Professor of Neurology at Tufts College Medical School, a former president of the American Neurological Association, a member of the Association of American Physicians and of the American Medical Association, and a psychopathologist of unique characteristics and marvelous accomplishment.

If you were to meet Dr. Prince at one of his numerous clubs, you would see in him a typical, courteous, highly cultured, self-contained Bostonian. You might be inclined to put him down as a man who had found life easy and taken it accordingly. Yet all his life he has been doing interesting things, strenuous things, big things. He is one of the most remarkably versatile men I have ever met. He is known in State Street as a successful manager of trust estates, in the hospitals of Boston as a physician who has labored tirelessly for the relief of suffering, among neurologists and psychologists as ranking in the very forefront of both professions, and by his fellow citizens generally as a resourceful, ardent, uncompromising civic reformer.

The city of Boston, indeed, owes more to Morton Prince than it can ever repay. He was the founder of the Public Franchise League, which of recent years has successfully waged two most important campaigns in behalf of the people against the gas and street railway companies. In the struggle of 1909 to secure the adoption of a new city charter, he took a leading part as chairman of the executive committee of the Committee of One Hundred.

In other respects also, Dr. Prince is a conspicuous figure in the life of Boston. Despite all the demands made on his time as man of affairs, physician, experimental scientist, and civic reformer, he has managed to keep up an active interest in athletics, dating from his college days at Harvard. He is an enthusiastic yachtsman, and was one of the founders of the old Myopia Hunt Club. But the sport which most strongly appeals to him — though he can no longer indulge in it — is football. And with right good reason, for it was he who, with H. R. Grant, introduced into Harvard, in 1874, the Rugby game out of which modern American football has since been evolved.

As a student, moreover, he distinguished himself for scholarship as well as for athletic ability. In his second year at the Harvard Medical School he won a Boylston Prize for an essay on " The Nature of Mind and Human Automatism," a paper of considerable significance as proof of the early age at which Dr. Prince took a serious interest in psychological problems. It was not until some years later, however, that he began to appreciate the importance and possibilities of medical psychology, and started in to experiment with hypnotism.

From that day — back in the early eighties — he has been continuing his explorations of the subconscious, with results that have enriched both psychology and medicine. He has been particularly successful in dealing with so-called " total dissociation of personality," a singular malady productive of the most intense mental suffering, but fortunately of comparatively rare occurrence.[1] Undoubtedly, though, his most helpful contribution to the development of scientific mental healing is found in the emphasis he has laid on the importance of what is known as " psychic reëducation."

Nowadays there is evident in psychothera-

---

[1] An account of a notable case of the kind will be found in Dr. Prince's "The Dissociation of a Personality." See also Boris Sidis's "Multiple Personality," and Isador Coriat's " Abnormal Psychology."

peutic circles a tendency to credit the origination of this valuable method to Dr. Paul Dubois, the well-known European neurologist. In reality the palm should be awarded to Dr. Prince, who was making use of psychic reëducation as early as 1890, and as long ago as 1898 published a detailed explanation of its principles and warmly advocated its use in the treatment of neurasthenia and other widely prevalent nervous disorders. Besides which, while Dr. Dubois seems to consider it the only effective method of mental healing, Dr. Prince recognizes that it is merely one of various methods, the choice of which depends on the character of the case in hand.

It is based on the discovery that nervous derangements can frequently be overcome by analyzing and explaining in the fullest detail to the patient the distinctly mental origin of his different symptoms, the circumstances giving rise to them, and the power which he himself possesses of throwing them off, the whole process thus being one of "reëducating" his reason and his will.

To illustrate, Dr. Prince once had a patient who came to him to be treated for neurasthenia characterized chiefly by extreme fatigue. She could not walk a block without becoming utterly exhausted. Patient inquiry traced the trouble to an unfortunate suggestion implanted in her mind by another physician who, when she first got into a run-down condition, had told her that she was suffering from lead poisoning. She had accepted and exaggerated this wrong diagnosis, and had subconsciously superimposed upon it the notion that she would inevitably be exhausted by the slightest exertion. In two weeks she was walking briskly, after Dr. Prince had made clear to her that the fatigue was a false fatigue, caused by self-suggestion.

A second patient, a woman thirty-five years old, had a morbid fear of fire. If a match were struck in her presence she would hunt everywhere, even in bureau drawers, for possible sparks that might cause a conflagration. Every night before retiring she spent an hour or more passing from room to room to make sure that there was nothing that could start a blaze. She was so afraid of fire that she could not be induced to go near an open fireplace where coal or wood was burning.

Inquiry showed that this abnormal dread had originated in a distressing experience she had had with fire many years before, and, having ascertained the origin of her "phobia," Dr. Prince was able to "educate" her into overcoming it.

It is, however, by no means always possible thus to argue nervous invalids into health. The method has distinct limitations, and must often be accompanied, or even superseded, by other psycho-therapeutic measures. Especially is this necessary when, as so frequently happens, the malady to be treated is rooted in emotional experiences of such remote occurrence as to be entirely forgotten by the victim. To recall these lost memories, Dr. Prince, unlike Dr. Dubois, freely avails himself of the remarkable power of hypnotism, or of the method of hypnoidization.

It remains to speak of the work of Dr. Sigmund Freud, of Vienna, a psychopathologist for whom his admirers advance the claim that he has " evolved not only a system of psychotherapy but a new psychology."

In one of Dr. Janet's cases, it will be remembered, a cure was obtained as soon as the emotional experiences responsible for the hysterical condition were recalled to the subject's memory. Freud — who, like Janet, studied under Charcot at the Salpêtrière — was much impressed by this and other cases similarly cured, and after his return from Paris to Vienna, in the early nineties, he began, in collaboration with another Viennese neurologist, Dr. Joseph Breuer, to treat hysterical patients by psychological processes.

His method was to hypnotize them, and

then question them about the origin of their symptoms, the effect being in many cases the disappearance of the symptoms as soon as the patient "worked off" the subconscious, forgotten emotion by recalling it and describing it to the psychopathologist.

But Freud found, as all psychopathologists have found, that it is not possible to hypnotize everybody, and that he would have to devise some other method applicable in the case of non-hypnotizable patients. The plan he ultimately hit upon was to urge and assure his patients that they could remember the facts he needed to get at, if only they would concentrate their attention and frankly tell him the thoughts, no matter how unpleasant, that came to them in connection with their symptoms. To this method he gave the name of "psycho-analysis."

I shall give but one case, typical of the many that have been treated successfully by him. It is that of an Englishwoman, employed as governess in the family of an Austrian manufacturer. The symptom of which she principally complained was a persistent hallucinatory odor of burnt pudding, which she seemed to smell everywhere she went. Close questioning by Dr. Freud traced the origin of this hallucination to an episode in the schoolroom when the children in her charge, affectionately playing with her, had neglected a little pudding they were cooking on the stove, and had allowed it to burn. But why this should cause the development of a hallucination was not at all obvious.

"You are, perhaps without knowing it, keeping something from me," Freud told her. "That incident distressed you greatly, or was connected with something else that distressed you. What was it?"

"I do not know," she said.

"Were you thinking of anything particular at the time?"

"Well," she replied, after much hesitancy, "I was thinking of giving up my position."

"Why?"

Gradually the truth came out. The governess had unconsciously fallen in love with her employer, a widower, whose children she had promised their dying mother to care for always. The episode of the burnt pudding represented a moment when some obscure scruple had urged her to leave the children because of something dimly felt to be wrong in her attitude of mind toward their father.

When this confession was made — a confession new to her as well as to Dr. Freud, for she had studiously concealed from herself her feelings with regard to her employer — the hallucinatory smell of burnt pudding disappeared. She had, by her avowal of the hidden truth, " worked off " the disease-producing emotion.

But, as the scent of the burnt pudding wore away, it became evident that another hallucinatory scent had underlain it and still persisted — the scent of cigar-smoke. Again Dr. Freud made use of his psycho-analytic method, and at length recalled to his patient's mind a scene which, while apparently trivial, afforded the correct explanation of the second hallucination. This scene she described to him as though it were a picture at which she was actually gazing.

"We are all sitting down to dinner, the gentlemen, the French governess, the children, and I. A guest is present, an old man, the head cashier. Now we are rising from the table. As the children leave the room the cashier makes as though to kiss them. The father jumps up, and calls out roughly, 'Don't kiss the children!' I feel a kind of stab in my heart. The gentlemen are smoking — they are smoking cigars."

Again, as Freud pointed out to her, there was an underlying emotional disturbance — the shock of discovering that the man she secretly loved could be so rough and harsh with another who was, like herself, one of his subordinates. She had tried to forget the incident, but it had remained a vivid memory in her subconsciousness, to produce in time the hallucinatory scent of tobacco, symbolical of the submerged memory. Like the smell of the burnt pudding the tobacco hallucination disappeared with her recital of the circumstances associated with it and she was enabled to recover her usual health and spirits.

In every case, Freud asserts, he discovered that, aside from the difficulty one would ordinarily experience in filling up memory-gaps, he had to overcome a considerable resistance on the patient's part, and that the resistance was due to the fact that the ideas to be remembered were all of a painful nature, of a character to give rise to feelings of shame, self-reproach, etc.

This led him to develop the theory that all hysterical and allied disorders are invariably the result of the repression of unpleasant ideas which one does not wish to remember. Probing still further, Freud found, as he believes, that the repressed ideas which were the immediate cause of the disease-symptoms were in their turn connected with other repressed ideas, often harking back to early childhood, and that these earlier ideas were, without exception, of a sexual character. On this basis he has built up an elaborate system of abnormal psychology, featuring the " instinct for reproduction " as playing the determining rôle in the development of hysteria, neurasthenia, and other nervous derangements.

Thus far, it must be said, scarcely another leading psychopathologist has accepted this sweeping, audacious theory. But it is being pressed vigorously by Freud and a rapidly increasing band of disciples, two of whom — Drs. A. A. Brill, of New York,[1] and Ernest Jones, of Toronto, Canada — have been ably presenting it for the consideration of American psychologists and physicians. By some Freud is regarded as having delved deeper than any other man into the mechanism of mentally caused diseases; by others he is condemned as an extremist who is " riding a hobby to death." Friends and opponents

[1] Under the title "Selected Papers on Hysteria and other Psychoneuroses," Dr. Brill has recently issued a translation from the writings of Freud embodying the latter's principal theories. It is published as No. 4 of the "Nervous and Mental Disease Monograph Series," in connection with *The Journal of Nervous and Mental Disease.*

agree, however, that, whatever his views, his psycho-analytic method of "tapping the subconscious" has resulted, like the method of hypnoidization, in placing a new and powerful instrument of diagnosis and therapy in the hands of the psychologically trained physician.

And that the physician of the future will also be a psychologist, there can be no doubt. The widespread interest manifest in medical circles, in medical institutions and periodicals, testifies abundantly to growing appreciation of the unquestionable truth that the labors of Janet, Freud, Prince, Sidis, and their fellow psychopathologists have opened a new era in the practice of medicine.

IV

Hypnotism as a Therapeutic Resource

A FRIEND once said to me, after reading an account of the tragic death of a traveling showman's "subject" while in a hypnotic trance:

"There ought to be a law in every State of the Union prohibiting the practice of hypnotism."

This has long been the opinion of many an intelligent and thoughtful person, yet to enact such a law would be not only a blunder, but a serious misfortune. Unquestionably, the giving of hypnotic exhibitions for mere amusement's sake, or the practice of hypnotism by amateurs for any reason whatsoever, ought to be strictly forbidden and heavily penalized. The gravest consequences may result from its use by the unskilled, as was only too clearly revealed by the tragedy just mentioned.

But prohibit the practice of hypnotism entirely, and mankind will be deprived of a great benefit conferred upon it by modern scientific discovery and investigation To-day it is known, as the result of many years of painstaking research by some of the world's foremost psychologists and physicians, that hypnotism is of far-reaching value in the treatment of many diseases.

Its efficacy, as has previously been pointed out, is due to the fact that the state of one's mind has a great deal to do with determining the health of one's body; that quite frequently diseases, even when their symptoms seem to be purely physical, have their origin in some mental disturbance, such as fright, grief, worry, anxiety, etc.; and that whenever this is the case they are curable by wholly mental means. Out of this discovery has been developed a new branch of the healing art — healing by suggestion.

It has been found that the effect of hypnotism is to increase marvelously the suggestibility of the hypnotized person. He responds with alacrity to suggestions which in the ordinary state would pass unheeded, and in this way it is often possible to bring about seemingly miraculous cures of disease. The neurasthenic, the hysteric, the semi-insane, the sufferer from so-called "functional" diseases involving perhaps paralysis of some organ of the body, the victim of the liquor habit and of moral defects and perversions — men and women who seem doomed to a life of hopeless invalidism, or to degenerate into social outcasts, are restored to health, and once more become useful and happy members of society.

In several countries — notably France, Germany, Holland, and Sweden — "hypnotic therapeutics" has won an established place in medical practice. In the United States this is not as yet the case, though hypnotism is used, either for diagnostic or therapeutic purposes, by many physicians of the highest standing. That it is not more generally used in American medical practice is due chiefly to wide-spread misconceptions concerning its true nature and the "way it works." These misconceptions are shared by the educated as well as the ignorant, by physicians as well as the laity, and are fostered by the extravagant claims of showmen hypnotists whose activities the law should curb.

Many persons refuse to be hypnotized because they are afraid it will "weaken their will." This is entirely fallacious, as is the notion that to be hypnotizable is of itself evidence that a person is of weak will. On the contrary, the more will power a person has, the more readily can he be hypnotized, for there are certain conditions involving the exercise of will power on the subject's part — for instance, concentration of the attention — that must be fulfilled before the hypnotic state can be brought on. For this reason, the weak-willed, the mentally defective, the insane, are very hard to hypnotize.

Dr. Voisin, a celebrated French alienist, found that he could not hypnotize more than ten per cent of the inmates of the asylum with which he was connected. Whereas an English experimenter named Vincent hypnotized with ease ninety-six per cent of a large group of university men.

These results are confirmed by the experience of American authorities. At the 1909 meeting of the American Therapeutic Association, Dr. Frederic H. Gerrish, who was then president, speaking from knowledge gained by much successful practice in hypnotic therapeutics, declared emphatically that nothing could be further from the truth than the prevalent belief that only the weak-minded, or, at best, the hysteric, are amenable to hypnotic suggestion. Said he:

"The experienced hypnotizer dislikes to deal with either of these classes of patients; he would rather for every reason have strong men with cultivated minds and disciplined wills. The physician who uses only physical therapeutic means prefers the well-balanced, sensible and intelligent for patients, and so does the man who employs psychic means, and for the same reasons. The hypnotizer asks his patient to exert his will in a specified direction; he wants the intelligent coöperation of the patient, and this requirement is most difficult for the feeble-minded, the untrained, the heedless, to meet."

In fact, it so frequently happens that hysterical and neurasthenic patients are in too agitated a condition to be hypnotized, that, as was stated on an earlier page, a leading American psychopathologist has invented a method to take the place of hypnotism and applicable in the case of patients who cannot or will not be hypnotized.

This brings us to another point concerning which there is wide-spread misapprehension: the fact that nobody can be hypnotized against his will. It is true that some people are so peculiarly constituted that they may be accidentally hypnotized merely by having a bright light suddenly flash in their eyes, or hearing a loud and unexpected noise. (These unfortunates form, as we shall see, an important exception to the general rule.) The average normal man or woman, however, need have no fear of being involuntarily thrown into the hypnotic state. The "magnetic influence" of the hypnotist, the power of the "hypnotic eye," of which one reads so often in current fiction and newspaper gossip, and which are even exploited in the modern drama, have no foundation in fact. The resolute opposition of one's will, together with refusal to comply with the necessary conditions, are quite sufficient to baffle any would-be hypnotist.

It is a great mistake, also, to suppose, as many people do, that a hypnotized person is entirely at the mercy of the hypnotist, even to the extent of committing crime at his bidding. Undoubtedly, however, erroneous though such a view is, much evidence might be cited seeming to confirm it.

For example, it is certain that a hypnotized person may be induced to perform the most fantastic, ridiculous, and unusual acts without offering the slightest opposition and apparently without any appreciation of the absurdity of what he is doing. Moreover, orders given to him in the hypnotic state, to be executed after he has been awakened, will be carried out at precisely the moment set for their performance, though it may be ten minutes, half an hour, a day, a week, or even months after the time he was hypnotized. On being awakened he has no recollection of what he has been told to do, yet when the appointed time comes he feels an irresistible impulse to execute the command given him during hypnosis.

Hundreds of experiments are recorded proving beyond the shadow of a doubt the binding force of "post-hypnotic suggestions," as these are called. Dr. Morton Prince, for instance, once was treating by hypnotic suggestion a patient suffering from insomnia, and having hypnotized her as usual, said:

"To-night you will go to bed at ten o'clock and will sleep soundly until seven to-morrow morning."

He did not know that his patient was that evening giving a large card-party. At precisely ten o'clock, although playing a hand of whist, she laid down her cards, asked to be excused, and left the room. The guests supposed she had merely gone upstairs for a few moments. But when ten minutes passed and she did not return, her husband went in search of her. He found her in bed, and fast asleep!

In one series of fifty-five experiments no definite date was fixed for the execution of the post-hypnotic command, the subject, a young woman of nineteen, being simply told that she was to perform a specified act at the expiration of a specified number of minutes, ranging from three hundred to more than twenty thousand.

Not once, on being awakened, did she remember what had been said, although she was offered a liberal gift if she could recall the commands given her. Nevertheless, of the fifty-five experiments only two were total failures, while in forty-five she executed the commands at exactly the moment designated, and in the remainder, was at no time more than five minutes out of the way. As to the complete failures the experimenter, Dr. J. Milne Bramwell, ascertained that in one instance she had entirely misunderstood the suggestion given to her, and in the other the circumstances were such that she might have carried out the command without his knowledge.[1]

It has even been found possible, through taking advantage of this peculiar phenomenon of post-hypnotic action, to employ hypnotic suggestion as a means of avoiding waste of energy in a responsible and exhausting occupation. Dr. Auguste Forel, in a paper contributed some years ago to a French scientific publication and translated in part by Frederic Myers, reports:

[1] A detailed report of these experiments is given in Dr. Bramwell's "Hypnotism," a book that may be recommended as a comprehensive and authoritative survey of its subject.

"At the Burghölzi Asylum, in order to watch the patients with suicidal tendencies during the night, we employ warders who have received appropriate hypnotic suggestions. The nurse's bed is placed at the side of the patient's, and the suggestion is given that she shall sleep well and hear nothing except any unusual sound the patient may make. If the latter attempts to get out of bed, or to do herself any harm, the nurse awakes at once; otherwise she sleeps soundly, despite the unimportant noises and movements made by the patient.

"This system succeeds admirably, provided we select suggestible warders for it. The appreciable advantage is that the nurse does not get tired (I have sometimes continued this sleeping watch with the same nurse for more than six months without her suffering the slightest fatigue), and that the danger of ordinary watching — that of falling asleep, despite every precaution — does not exist. I have not had a single accident to report, with regard to patients watched in this manner, for four years. It is curious to see the surprise sustained by the said patients — melancholics — to see themselves so well watched in this way."

And, to cite one more illustration, emphasizing both the compelling quality of post-hypnotic commands and the absence of all knowledge of them by the subject when in the waking state, let us take an experience reported by that pioneer French suggestionist, Dr. Hippolyte Bernheim. Dr. Bernheim once hypnotized a soldier, and asked him:

"On what day in the first week of October will you be off duty?"

"On the Wednesday."

"Well," said Dr. Bernheim, "on that day you will pay a visit to Dr. Liébeault; you will find in his office the President of the Republic, who will present you with a medal and a pension."

The soldier was then awakened and closely questioned as to what had been said to him, but could remember nothing, as his answers showed conclusively. However, on Wednesday, October 3, Dr. Liébeault wrote to Dr. Bernheim:

"Your soldier has just called at my house. He walked to my bookcase, and made a respectful salute; then I heard him utter the words, 'Your Excellency!' Soon he held out his right hand, and said 'Thanks, Your Excellency.' I asked him to whom he was speaking. 'Why, to the President of the Republic.' He turned again to the bookcase and saluted, then went away. The witnesses to the scene naturally asked me what that madman was doing. I answered that he was not mad, but as reasonable as they or I, only another person was acting in him."

Facts like these — the passive, unresisting obedience of the subject, his readiness to carry out the most absurd orders long after the time he was hypnotized, and his total ignorance in the waking state of all that had been said to him during hypnosis — naturally gave rise to the suspicion that there might be hypnotic crimes; that unscrupulous men might make innocent persons an unconscious tool of the basest designs.

To test the truth or falsity of this theory, Dr. Liégeois, another French investigator, undertook a number of elaborate experiments, with results which astounded him.

One subject made a fictitious will in his favor, another signed a note for money never received; a third was induced to dissolve in a glass of water a powder which he had been told was arsenic, and to give it to his aunt to drink; a fourth discharged a revolver point-blank at a witness to the experiments. Not one of the subjects, after being awakened, realized what he or she had done.

Liégeois, appalled by what he considered conclusive proof that hypnotism might be used for criminal purposes, hastened to publish a report [1] of his experiments. This re-

[1] "La Suggestion Hypnotique dans ses Rapports avec le Droit Civil et le Droit Criminel." Published in 1884, and five years later expanded into a treatise, "De la Suggestion et du Somnambulisme dans leurs Rapports avec la Jurisprudence et la Médicine Légale."

port was widely copied, quoted from and commented on, and to the profound impression it made is largely attributable the present erroneous state of public opinion regarding hypnotism and crime.

For, despite these amazing experiments, there is in reality little likelihood of crimes being committed under hypnotic influence.

Hardly had Liégeois' report appeared than critics pointed out that quite possibly his subjects were " subconsciously " aware that they were only " play acting "; and that if real crimes had been suggested to them, they would have refused compliance just as quickly and firmly as they would have done in their normal state. To test the correctness of this theory other experiments were tried, and an overwhelming mass of evidence was soon accumulated controverting Liégeois' views.

In one instance the hypnotized subject readily consented to stab a man with a bit of cardboard which she had been told was a dagger. But she absolutely refused to carry out the order when an open pocket-knife was placed in her hand. The experimenter says:

"I have tried similar experiments upon thirty or forty people, with similar results."

Often, when a suggestion of real crime is given, the subject at once awakes from the hypnotic sleep. And in some cases evidence has been obtained showing unmistakably that the reason subjects refuse to obey such commands is because they are repugnant to their moral sense.

Thus, a young woman whom Dr. Bramwell had often hypnotized, and who was an exceptionally docile subject, was told by him that, on awaking, she was to steal a watch which a mutual acquaintance had left on a table.

"He is very absent-minded," said Dr. Bramwell, "and will never miss it. Or, if he does, he will not remember where he left it. So you may take it safely enough. You know you need a watch badly."

He then brought her out of the hypnotic

state, but she made no effort to take the watch, and indeed, did not even seem to notice it. Finally, he asked her:

" Don't you see that watch on the table? "

" Yes."

" Do you know whose it is? "

" Yes."

" Don't you feel like taking it? "

" Of course not."

She was rehypnotized, and questioned anew. Yes, she said, she remembered quite well that she had been told to steal the watch.

" Why did you not do so? " asked Dr. Bramwell. " Was it because you were afraid of being found out? "

" Not at all. It was because I knew it would be wrong."

In fact, no competent exponent of hypnotism to-day believes that a person is inevitably obliged to execute all hypnotic commands given him. And while some still cling to the idea that hypnotic crimes are possible, the consensus of scientific opinion is that no person who would not in his normal state perpetrate the crime suggested, would perpetrate it if hypnotized.

It is equally certain, though, that under hypnotic influence people are liable to accuse themselves of crimes they have not committed. This is a real danger, which ought to be carefully guarded against in courts of justice.

There is reason to believe that many "police confessions," extorted from accused persons by the processes of the so-called "third degree" and afterward found to be untrue, are made in a hypnotic state. The persistent questioning of the prisoner by the police, their pitiless insistence that "he is guilty and knows he is guilty," may develop in him that peculiar hysterical condition in which, as has already been said, he may become spontaneously hypnotized by an unexpected noise or the sudden flashing of a light.

It is not so long ago that a young man, scarcely more than a boy, was executed in

Chicago for a brutal murder, having been convicted on his own confession made to the police. Six days before his execution he suddenly asserted his innocence, and declared that he had not the least recollection of having made any confession. He could only remember that while he was being questioned a revolver was pointed at him. "I saw the flash of steel, and after that everything is a blank to me." There is warrant for suspecting — as one well-known psychologist, Professor Münsterberg, in his book, "On the Witness Stand," has boldly said was actually the case — that the sight of the shining revolver barrel plunged the terrified youth into a state of spontaneous hypnotism, in which he readily accused himself of the crime his inquisitors charged him with having committed.

The gravest danger of hypnotism has still to be mentioned. If one need not fear that it will weaken his moral fiber, place him absolutely in the power of the hypnotist, or make him an unconscious instrument of crime, there is none the less an important reason why no one should allow himself to be hypnotized by any but a recognized expert. In the hands of an amateur a person not physically fit to be hypnotized may collapse during the trance condition, as in the case of the tragedy to which allusion was made above; or a violent attack of hysteria may be provoked instead of the quiet, refreshing sleep of the true hypnotic state.

In one instance recorded in medical annals, a young man at an evening party consented to be hypnotized "for fun" by an operator who knew no more about hypnotism than his subject did. Two attempts to entrance him failed completely. At the third he began to shiver, fell to the floor, jumped up, laughed, sang, wept and acted as though demented. After a time he became cataleptic, and for ten days had a succession of convulsive attacks, loss of speech and catalepsy. It was more than three weeks before he was entirely normal again.

Mishaps like this are unknown in the experience of the qualified medical hypnotist. Dr. Liébeault, a most conscientious man, after thirty years of practice, during which he hypnotized thousands of persons, stated that he could not recall a single case in which he regretted having used hypnotism. Professor Forel, another eminent authority, testifies that " Liébeault, Bernheim, Wetterstrand, van Eeden, de Jong, Moll and I myself declare categorically that although we have seen many thousands of hypnotized persons, we have never observed a case of mental or bodily harm caused by hypnosis, but, on the contrary, have seen many cases of disease cured or relieved by it."

One need not hesitate, therefore, in placing himself in the hands of a competent, conscientious practitioner if one is suffering from a malady that hypnotism might benefit. It should be kept clearly in mind, though, that the ordinary physician, however conscientious he may be, is not qualified to practice hypnotism unless he has had a psychological as well as a medical training.

Hypnotism is so essentially a matter of the mind, and its successful operation depends so entirely on knowledge of how the mind works, that the trustworthy hypnotist must be a psychologist as well as a physician. For this reason only men of standing and experience should be consulted.

V

Secondary Selves

THAT dreams may come true, that fact often is stranger than fiction, that the word "impossible" should almost be banished from our vocabulary, has perhaps never been more impressively demonstrated than in the remarkable parallel between the happenings of real life and the fancies of imagination as exemplified in that wonderful masterpiece of story-telling — Robert Louis Stevenson's "Dr. Jekyll and Mr. Hyde."

Stevenson, as he himself has told us in one of his delightful essays,[1] obtained the plots for many of his stories from dreams that came to him in the quiet of the night. "Dr. Jekyll and Mr. Hyde" was one of these dream stories. So vivid, so terrifying

[1] "A Chapter on Dreams," in the volume "Across the Plains."

was the impression it made on him that in his sleep he uttered cries of horror, until his wife awoke him in alarm. Next morning, with the memory of the dream fresh in his mind, he began to set it down on paper, and, though weakened by illness, in less than a week had ready for the printer the tale that has thrilled the hundreds of thousands who have read it or seen it acted on the stage.

Thrilled them — but left them firmly persuaded of its absolute incredibility. In real life, they have felt, such things cannot be. No man can be two persons, two selves, two individualities — the one self, like Dr. Jekyll, leading a wholly blameless life and being universally respected and beloved, while the other self, Mr. Hyde, is a monster of iniquity, revelling in vice and crime. No man can, as the Dr. Jekyll of Stevenson's dream story did, drop in a moment his benignant and noble self and be transformed into the malevolent self of Mr. Hyde, seesawing back and forth between the two selves to the day of his death. This is all very well in a story, but it is not true to life. It is quite impossible.

Yet in reality it is so far from being impossible that within recent years it has been proved beyond the slightest doubt, as the result of careful investigation by eminent scientists, that the strange case of Dr. Jekyll and Mr. Hyde frequently finds a counterpart in the lives of real men and women. To-day, appalled by the discoveries that have been made, psychological science, medical science, and legal science, are desperately striving to ascertain just what ought to be done with the Dr. Jekylls and Mr. Hydes of actual existence — are trying to learn why it is that they become two persons instead of one, how they should be treated to prevent the appearance of the baser self, and how far they should be held accountable for their actions.

This last problem is of the utmost urgency because — unlike the hero of Stevenson's fanciful tale, and making the situation far more terrible and tragic than even Stevenson conceived it — the Dr. Jekylls of real life have no knowledge whatever of what they say and do while they are Mr. Hyde. And, on the other hand, though this is not always the case, when they are Mr. Hyde they know nothing of themselves as Dr. Jekyll. There is a complete cleavage, a blotting out of all memory, between the two selves.

A few years ago there lived in a Western city a gentleman whose identity I shall, for reasons that will become obvious, conceal under the assumed name of Mr. Brown. He had had a successful business career, and was at that time the president of a local bank, being regarded as one of the most responsible and trustworthy members of the community. His home life was all that could be desired. Married at an early age, he was still as devotedly attached to his wife as in the days when he went courting her. He had one child, of whom he was also passionately fond. Successful, wealthy, happy in his domestic relations, he was in every way a man to be envied.

One morning he left home intending, before going to the bank, to ride out into the country and collect some rent that was due him. On the way he suddenly felt strangely dizzy, and remembered nothing more until, a couple of hours later, he walked into the bank. Questioned by his partners he could give no account of his movements. What had he done with his horse? He did not know. Had he collected the rent? He did not know. All he could remember was feeling dizzy, dismounting, and standing in a doorway.

But when he looked in his pocket-book he found that it contained the exact amount of the rent he had set out to collect. A messenger was hurriedly sent to make inquiries.

"Why," he was told by the woman of the house, "of course Mr. Brown was here this morning. I paid him myself, and here is his receipt to show for it. But I must say that I hardly knew him when he came in. He looked awful. He looked as though he would like to kill me. I was terribly frightened, and was mighty glad when he took the money, signed the receipt, and went away."

Though greatly disturbed by the messenger's report, Mr. Brown decided to say nothing to his wife about this singular adventure. It would, he thought, only alarm her needlessly, for surely nothing like it would occur to him again.

Exactly a year later, while in his office, the same feeling of dizziness came upon him. He did not fall, and it lasted only a moment. But it had the amazing effect of temporarily depriving him of all knowledge of his business and family relationships, and even of his own identity.

Addressed by name he made no response; when his partners anxiously asked him if he did not know them, he calmly replied that he most decidedly did not. And, precisely like the Mr. Hyde of Robert Louis Stevenson's weird tale, there was something in his manner and in the expression of his eye, that warned them it would be dangerous to oppose him. To their intense relief, at the end of two hours he seemed to be completely himself once more. But he insisted that he knew nothing of what had happened in the interval.

Another year passed, and again the dizzy sensation overpowered Mr. Brown, this time with disastrous consequences.

It was his custom every morning to kiss his wife and child good-by, when starting downtown. He had done this as usual one day — waving, indeed, a farewell to them until he was out of sight, and seeming to be in the best of health and spirits.

Ten minutes after his departure Mrs. Brown was astonished to hear the front door opened violently by somebody who ran hurriedly from room to room through the lower part of the house. While she listened with increasing amazement and alarm, hasty steps sounded on the stairs, the door of her bedroom was flung open, and her husband rushed in.

She scarcely recognized him. The eyes

that had been gazing lovingly into hers so short a time before, were wild and staring. The lips that had kissed hers were set in a thin, hard, cruel line. His whole face was distorted in a frenzy of ungovernable anger and hatred.

"John, John!" she gasped. "What is the matter?"

"Oh, you are here, eh?" he cried. "Curse you! I have been looking everywhere for you."

Without another word he leaped forward, seized her by the throat, and began to choke her.

Back and forth across the room they struggled — she fighting desperately for life, he intent on killing her. Suddenly, when she had almost reached the limit of her resistance, he uttered a horrible shriek and threw her from him, himself falling on the bed, where he immediately lapsed into unconsciousness.

Too weak even to call for assistance, Mrs. Brown lay on the floor, half-leaning against a chair, until, an hour later, she was roused by her husband's voice, anxiously calling:

"Why are you lying there, Mary? What has happened. I have had such a fearful dream. Or was it a dream? Have I harmed you? Have I harmed you?"

Rising feebly, she showed him the finger-marks on her throat. With tears streaming down his face, he assured her that he had not the least knowledge of the murderous assault. Everything was a blank to him. The last thing he remembered was a sudden feeling of dizziness while walking to the office.

A month afterwards he again attacked her, while she was attending to their child, who had been stricken with what proved to be a fatal illness. Dragging the child out of her arms, he flung it roughly aside, and seized her by the throat as before. But this time she was able to summon aid.

The doctors who were called in talked sagely of "overwork" and "nerve strain," and advised the unhappy banker to leave home and take a complete rest. While away he wrote almost daily to his wife, his letters breathing the profoundest devotion and love. But she noticed that there were periods, of a week or two at a time, when he seemed to be quite unaware of existing conditions. During these periods his letters made no reference to their dead child, but were almost duplicates of the letters he used to write to her before their marriage, so much so that it well-nigh seemed that time had turned backward for him, and that he was living again in the days of his youth.

He returned home a week before Christmas. All went well until Christmas Eve, when, having gone downtown to make some purchases, he came back carrying a few cheap presents in one hand and a rifle in the other. Leveling the rifle at his terrified wife, he compelled her to promise him that within a week she would leave the house never to return to it. And, in fact, fearing for her life, she next day left him for evermore.

All this time he had managed to conduct his business affairs as shrewdly as ever; and with the departure of his wife all murderous impulse seemed to leave him. But the periods of his strange forgetfulness now became more frequent and more prolonged. He became, as it were, two men, leading separate lives, and with separate chains of memory for the events of each existence, neither of his two "selves" having any knowledge of what the other did. In despair, he finally consulted a physician of Kansas City, Dr. S. Grover Burnett.

"He confided to me," says Dr. Burnett, "that he had domestic trouble. He spoke of his intense love for his wife, a love that had firmly cemented their lives from early adolescence, and had grown with age and maturity till he had long learned to look upon her as that part of his life without which there was a sunless future not worth the living.

"Something, he told me, had come into his life that made him dangerous to her. What it was he did not know, but from his periods of memory blanks and from certain confused and incomplete memories that hung over him with a blurred horror, and the details of these periods as told him by reliable persons, he realized that his wife was justified in fearing him. He was perfectly conscious of the cheerless future held out to him, and as he sat intelligently portraying it to me, he was a picture of pathos — a strong man weeping like a child."

Dr. Burnett questioned him closely.

"You have just returned from a six weeks' trip in the South, have you not?" he asked. "What happened there?"

"I only know that I have been South from what my friends tell me," he replied, "and from certain entries in this book." And he drew from his pocket a large memorandum-book.

"What is that book for?" inquired Dr. Burnett, wonderingly.

"To enable me to keep track of myself, and carry on my business. I jot down in it names and dates and details of conversation, so that the next time I meet a man, if I fail to recognize him I have only to consult my notes to get the matter straightened out. In this way I prevent anyone from suspecting the true state of affairs. Nobody, except my closest friends, even suspects that I am two persons instead of one. But," he added, bitterly, " my life is ruined."

As, indeed, it was. His wife, abandoning all hope that it would ever be safe to return to him, sued for and obtained a divorce; the publicity attending the court proceedings impelled him to sell out his business and remove to another State; and when last heard from he was still oscillating between the two personalities, and still depending on the memorandum-book to keep his secret hidden from the outside world.[1]

When the Mr. Hydes of real life develop murderous tendencies, those nearest and dearest to them are almost invariably the ones in greatest danger. It was thus, as

[1] Dr. Burnett's report of this case will be found in *The Medical Herald,* vol. xxii.

Secondary Selves

we have just seen, in the case of the western banker, and it was thus in another singular case, occurring in New York City.

Among the residents of a crowded East Side tenement was a respectable German-American family of five persons — father, mother, and three children, the oldest child a boy of six, the youngest a mere infant. The father was a hard-working, temperate, thrifty man, who, in the nine years of his married life, had been a model husband. Although in the humblest circumstances, and often hard pressed to make both ends meet, he and his wife were thoroughly happy in their love for each other and for their children.

It was his habit upon returning from work to pass half an hour or so playing with the children before they were put to bed. One evening, after romping with them as usual, he gave each of the little ones a kiss, put on his hat, and went out for a walk, telling his wife that he would not be long gone. As she afterwards testified, there was nothing strange or alarming in his manner. He seemed entirely calm and rational — in every way his light-hearted, genial, kindly self.

But in the ten minutes that he was absent from the apartment something extraordinary happened to him — just what remains a mystery to this day. He came back with all the love for his wife and children driven from his heart, and in its stead a blind, unreasoning hatred. Gazing fiercely around the room, he suddenly pounced on his second child, lifted him up, and hurled him bodily through the open window. Another instant and he had similarly seized his first-born in an iron grip and had thrown him too out of the window, after which he turned in swift pursuit of his wife, who had snatched up the baby and fled shrieking down the stairs.

As luck would have it, a fire escape ran past the window through which the children had been thrown, so that neither of them fell to the street, or was seriously hurt; and before their frenzied father could attack his wife he was overpowered by neighbors and held until the police came and took him to the psychopathic ward of Bellevue Hospital, where he was given an opiate and put to bed, the physician in charge fully expecting that when he awoke in the morning it would be as a raving maniac.

Instead, he awoke in perfect possession of his senses, but with utter oblivion to the happenings of the previous night. Excitably he demanded where he was, and why he had been brought there, and listened horrified and incredulous when told what he had done. His wife was summoned, but not even she could convince him that he had actually tried to murder his family. Pitifully he begged to be allowed to return home and go to work. Nor did there seem to be any valid reason for detaining him, for he clearly was not insane in the ordinary sense of the term.

Still, for safety's sake it was decided to keep him under observation, and he was accordingly committed to an asylum. Time justified this precautionary measure. Six months after his committal, it was observed that he seemed to lose all awareness of his surroundings. His faculties were unimpaired, but he acted as though the physicians, the attendants, and his fellow patients were total strangers to him. This lasted several days, and culminated in a violent outbreak during which he severely injured an attendant. He then fell into a deep sleep, out of which he awoke with his memory a blank for the events of the previous few days. The same thing occurred six months later.

The true nature of his trouble was now plainly evident. He was a man with two selves — a Dr. Jekyll self, good, kind, lovable, and a Mr. Hyde self, so malignant and cruel-minded that it would be a crime against society to allow him his freedom.[1]

Fortunately for themselves and for those around them the people with two selves by no means always display such radical alterations of character and conduct. What often

---

[1] This case is reported by W. J. Furness and B. R. Kennon in the *New York State Hospitals Bulletin*, vol. ii.

happens is that when they develop the second self they merely — though this is sad and terrible enough — lose all memory for the events of their previous existence. Sometimes their memory loss is so complete that they may literally be said to be born into the world a second time. They have forgotten all the acquisitions of education and experience, and, like any child, have to be taught to read and write, and occasionally even have to learn how to talk and walk. An interesting case of this sort is reported by Dr. Charles L. Dana,[1] the celebrated neurologist, Professor of Nervous Diseases in Cornell University Medical School.

His patient was a young man of twenty-four who, up to the time of his peculiar "accident," had never betrayed any sign of abnormality. He was but recently out of college, was employed in a large business house, and was engaged to be married to a beautiful girl whom he had known almost

[1] Another striking case of the same type will be found in my earlier book, "The Riddle of Personality." Dr. Dana reports his case in *The Psychological Review*, vol. i.

since his boyhood. Shortly before the day set for their wedding he met with a financial reverse that worried him greatly. He lost his appetite, could not sleep, and in many other ways showed that he was laboring under an intense mental strain. One night, soon after going to bed, he was seized with a nervous chill which lasted some hours and was followed by a deep, stuporous sleep. After regaining consciousness he acted so strangely that his relatives feared he had gone insane.

When spoken to, he answered in a strange, hesitating, almost unintelligible way, and showed clearly that he did not know who he was, or in what relation he stood to his father and mother, his brother and sister. His fiancée was hastily sent for, but he greeted her with a vacant stare in which there was no recognition. A letter was handed to him. He could not read it.

The mystery was solved as soon as Dr. Dana was called in consultation. "Your son," he told the agonized parents, "is not insane. He has simply lost his personality. You will have to educate him over again, and trust to time to bring about his complete recovery."

It quickly developed that, although essentially a child, he could learn things far more readily than any child. In a few weeks he could read and write almost as well as ever, thanks chiefly to the efforts of his grief-stricken sweetheart, who devoted hours every day to his instruction. He also acquired a remarkable shrewdness that enabled him to conceal from outsiders his anomalous condition. But he maintained his attitude of childlike wonderment at everything around him.

"If one were to meet him, and discuss ordinary topics," says Dr. Dana, who saw him constantly, "he would show no evidence of being other than a normal man, except that he might betray some ignorance of the nature or uses of certain things. His conversation ran chiefly on the things he did every day and on the new things he every day heard. He was exactly like a person with an active brain set down in a new world, with everything to learn. The moon, the stars, the animals, his friends, all were mysteries which he impatiently hastened to solve."

Three months after the onset of his trouble, his real personality returned as unexpectedly and mysteriously as it had vanished. His brother had taken him to spend the evening with his fiancée, and he had seemed to her on that occasion more unlike himself than ever. After his departure she had cried bitterly, feeling that there was no hope he would ever get well. On the way home, he complained to his brother that his head felt prickly and numb, and he had no sooner got into the house than he fell asleep and was carried upstairs and put to bed.

An hour later he awoke with his memory for the past completely restored — the past, that is to say, up to the moment of his nervous chill. Of the happenings of the three

Secondary Selves   145

months that had since elapsed he knew nothing, absolutely nothing. And to this moment he remembers nothing of them. A quarter of a year has dropped out of his life, which he can account for only by what his friends tell him.

Usually, however, the development of a second self does not involve such far-reaching loss of knowledge, such a startling reversion to the state of infancy, as occurred in the case of this unfortunate young man. As a general thing the sufferer simply loses all sense of his identity, kindred, and true position in the world. He retains his intellectual faculties unimpaired, gives himself a new name, and takes a fresh start in life, often earning a comfortable living, sometimes amassing wealth, and troubled only by the fact that he can remember nothing of his early history.

A case in point is that of a Rhode Island clergyman who disappeared from the city of Providence under circumstances that led his family to fear he had met with foul play.

What had actually occurred was that, while riding in a street-car between Providence and Pawtucket, a profound psychical change took place in him, completely erasing from his memory all knowledge of his previous life.

He found in his pocket a large sum of money which he had happened to draw from his bank that morning, but nothing to indicate his identity or place of residence. However, this seems not to have troubled him in the least. He took a train for New York, registered at a hotel under the first name that came into his head, and, after spending some days in New York and Philadelphia, finally wound up at a little town in Pennsylvania, where he set up for himself as a storekeeper.

Several weeks later his true self returned to him, and, as may be imagined, he was greatly surprised and frightened to find himself in a strange town, masquerading under a name not his own, and working behind a counter instead of in the pulpit. A couple of telegrams straightened matters out, and he soon was once more with the relatives who had been mourning him as dead.[1]

In another case, a Philadelphia grocer, while out securing orders, stopped two boys in the street and asked them if they would take his team back to the address painted on the wagon. Then he walked briskly away, and nothing was heard of him for a month, when he appeared at his home in a dazed, emaciated condition, his clothing in rags, his boots almost worn off his feet. It was subsequently learned that he had been leading the life of a tramp, wandering from town to town, working at whatever jobs he could get, and utterly ignorant of his true identity until self-consciousness at last returned to him while he was walking a railroad track near Baltimore.

A Petersburg, Va., business man had a very similar experience. After a two days' visit to New York, during which he did a

[1] This is the classic Ansel Bourne case, studied by Professor James and Dr. Hodgson, and reported in detail in the *Proceedings of the Society for Psychical Research*, vol. vii.

great deal of buying for his firm, met many acquaintances, and exhibited no indications of mental aberration, he started for home by steamer. When the tickets were collected he was missing. No one had seen him leave the boat, jump or fall overboard. All manner of theories were advanced to account for his disappearance, and a vigorous search was made for him, but to no avail.

Six months afterwards, when all hope had been given up and the court had appointed an administrator for his estate and a guardian for his children, word was received that he had suddenly appeared at a relative's house in a distant Southern city. He was wearing the same suit of clothes that he had on at the time of his disappearance, but was so changed physically that he was almost unrecognizable. He had lost a hundred pounds in weight, and was extremely feeble. Here is his own account of what had happened to him:

"I was feeling very tired — thoroughly

fatigued — after a very busy day in the city, so went to my stateroom immediately upon going aboard the boat and changed my clothes. Up to that time I was thoroughly conscious, but after that I recall nothing — all was oblivion — till six months later when I came suddenly to myself in a distant city in the South, where I knew no one.

"I found myself driving a fruit-wagon on the street. I was utterly astounded. Why I was there, how and when I got there, where I came from, what I had been doing, were puzzling questions to me. Upon inquiry I learned that I had been there, and at work, for some time. My life since I was in that stateroom had been an absolute blank to me. I can give no account of myself during that period of time. I started at once for Virginia, but on the way I again lost consciousness, though only for a day or two. When further on my way home, I felt so utterly worn out, I stopped in a certain town and went to the house of a very near relative. From there I was taken home.

I was in a half-dazed, confused condition, and remained so some days longer." [1]

In all of these cases the appearance of the second self and the disappearance of the true self, was a temporary, transient affair, lasting not longer than six months. But there is evidence indicating that the change may be permanent, and that the original personality may never return.

There was brought to a Portland, Ore., hospital a young man who had been badly hurt by falling from a barge and striking his head against a log. He was delirious for several days, and his life was despaired of, but suddenly his mind cleared up and he seemed to be in as perfect health as before the accident.

The hospital physicians were surprised to find, however, that he had no recollection of having been injured, and, in fact, had forgotten all about the occurrences of his life for the previous four years. He spoke of having had a quarrel with his father " yes-

---

[1] W. F. Drewry's "Duplex Personality," in *The Medical News*, vol. lxviii.

terday," and when asked what he meant by "yesterday" gave a date in the year 1898. It was then 1902. He expressed great amazement when told he was in Portland, asked whether it was Portland, Ore., or Portland, Me., and said he knew nothing of how he got there, but supposed that he was still in his home town of Glenrock, Wyo. From a friend with whom he had been living in Portland, it was learned that he had never said anything about his past, and often acted "queer."

There happened to be in the hospital a physician, Dr. J. Allen Gilbert, who knew that when a person was hypnotized he could often recall memories which had faded from his consciousness, and it was decided to try the effect of hypnotism on this puzzling patient. Put into the hypnotic state by Dr. Gilbert, he was able to give a full account of his lost four years.[1] And a most surprising story he unfolded.

[1] This will be found in more detail in Dr. Gilbert's account of the case, as contained in his "A Case of Multiple Personality," in *The Medical Record*, vol. lxii.

After the quarrel with his father, which had ended in the latter's hitting him over the head with a shovel, he had run away from home, enlisted for the Spanish War, and accompanied his regiment to Chickamauga. There he had fallen ill, deserted, and secured work from a farmer at Green Brier, Tenn. But he soon got tired of farm-work and started West, tramping it from one town to another, until at last he reached San Francisco, where he again enlisted and again deserted, becoming a fireman on a steamer plying between San Francisco and Portland. He had settled permanently in the latter city in August, 1901.

Unfortunately, the moment he was dehypnotized all this knowledge again slipped from his memory. He could remember nothing from the moment of the quarrel in 1898, and was considerably surprised when the military authorities of the nearest post placed him under arrest on a charge of desertion from the United States army. He was, of course, acquitted after the true inwardness of the case had been explained, and in the end Dr. Gilbert, by the aid of hypnotism, was able to bring about a return of memory for the facts of his entire life, and fuse the two selves into one.

Had it not been for the fall from the barge it seems altogether likely, though, that this young man would have remained to the day of his death in ignorance of his real identity and early history.

Nowadays, indeed, hypnotism is frequently used by specialists in the treatment of mental and nervous diseases, as a means of curing this strange and dreadful malady of the "double self."[1] It cannot be too clearly appreciated, of course, that hypnotism should never be used for any purpose whatsoever by any but thoroughly qualified experts, for it is a dangerous agency in the hands of the unskilled. Nor, even when employed by experts, is it always successful in

[1] Notably in the case of Miss Christine L. Beauchamp, described by Dr. Prince in his book "The Dissociation of a Personality."

curing victims of double personality, because, as investigation has made certain, the appearance of the second self is sometimes due to deep-seated, irremediable changes in the physical organism, which no known remedy can reach.

But when the cause is purely psychical — the result of worry, nerve strain, etc., as in the case of the young man who was born again — hypnotism can always be utilized by the trained practitioner with hope of good results; and it frequently is effective, as the case of Dr. Gilbert's patient shows, when the cause is not primarily psychical but physical, such as a blow on the head.

Moreover, the investigation of cases like those I have described, has led within the past few years to the discovery that quite often people suffer not from a total but from a partial disintegration of personality, taking the form of certain hitherto very baffling diseases — hysteria, neurasthenia, psychasthenia. When this is the case hypnotic suggestion — and even suggestion in the waking state — can also be employed to bring about a cure.

Truly, therefore, it may be said that we are just beginning to comprehend the complexities and intricacies of our psychical makeup, our mind, our personality. Truly we are living in a wonderful age — an age of great discoveries, of marvelous promise for the future, when science, by methods which it is gradually evolving by laborious experiment, shall develop our inner resources to an extent formerly undreamed of, shall give greater potency and stability to our personality, and shall make real Dr. Jekylls and Mr. Hydes almost a thing unknown.

VI

Psychology and Everyday Life

IN the preceding pages the effort has been made to give some idea of the remarkable contributions made by modern psychology to the practice of medicine — contributions of such profound significance that to-day many specialists in nervous and mental diseases are successfully treating their patients by psychological rather than medical methods, while the general practitioner also is in many cases using psychological knowledge to reinforce the curative value of ordinary therapy. But the helpfulness of modern psychology is by no means confined to the physician. It equally proffers aid to the parent, the educator, the sociologist, the criminologist, the lawyer, the judge, the manufacturer, merchant, and artisan, the writer, public speaker, artist, and musician. In fact, it is not too much to say that there is no field of human endeavor in which benefit may not be had through wise application of the discoveries of psychological research.

Only a comparatively short time ago, it is true, this could not be said. As late as the seventies of the past century, psychology was regarded, and not without reason, as one of the most impractical of sciences, of philosophical and theoretical importance, no doubt, but incomparably inferior to physics, chemistry, geometry, and other branches of science with reference to the possibility of its finding practical application. All this was changed with the establishment by Professor Wundt at Leipzig University of the first laboratory for experimental psychology. Wundt and his pupils, and other experimenters in various countries, invented and perfected apparatus and methods for investigating the processes of the human mind with a precision impossible to earlier psychologists. Such instruments as the chronoscope for measuring, even to thousandths of a second, the rapidity of thought, the sphygmograph for studying the emotions, and the ergograph for ascertaining the exact characteristics and consequences of fatigue, together with the discovery of the tremendous value of hypnotism and hypnoidization as means of getting at subconscious mental states, have enabled psychologists to make more progress during the past thirty years than throughout the previous two thousand years of the history of psychology.

Of course, as an applied science psychology is still in its infancy, and holds more of promise to mankind than of actual achievement. Yet it would be difficult to name another branch of science which has accomplished so much in an equally short time after it was first put on an experimental basis. Outside of the field of medicine, in which thus far psychology has proved itself most conspicuously useful, it is now being applied with striking results in such diverse fields as education, social reform, law, commerce, and industry. Of its utilization as an aid in education a most impressive example is found in the work carried on at the psychological clinic of the University of Pennsylvania in behalf of the mentally retarded children of Philadelphia. The clinic was first established in 1896, and its origin and development are interestingly described in the following statement by its director, Professor Lightner Witmer:

" The occasion was given for the inception of this work by a public school teacher, who brought to the psychological laboratory of the University of Pennsylvania a boy fourteen years of age for advice concerning the best methods of teaching him, in view of his chronic bad spelling. Her assumption was that psychology should be able to discover the cause of his deficiency and advise the means of removing it. Up to that time I could not find that the science of psychology had ever addressed itself to the ascertainment of the causes and treatment of a deficiency in spelling; yet this is a simple developmental defect of memory, and memory is a mental process concerning which the science of psychology is supposed to furnish authoritative information. It appeared to me that if psychology was worth anything to me or to others, it should be able to assist the efforts of the teacher in a retarded case of this kind.

" The absence of any principles to guide me made it necessary to apply myself directly to the study of the mental and physical condition of this child, working out my methods as I went along. I discovered that the important factor in producing bad spelling in this case was an eye defect. After this defect had been corrected, his teacher and I worked together to instruct him as one would a mere beginner in the art of spelling and reading. In the spring of 1896, when this case was brought to me, I saw several other cases of children suffering from the retardation of some special function, like that of spelling, or from general retardation, and I undertook the training of these children for a certain number of hours each week. Since that time the laboratory of psychology has been open for the examination of children who have come chiefly from the public schools of Philadelphia and adjacent cities. The University of Pennsylvania thus opened an educational dispensary. It is in effect a laboratory of applied psychology, maintained since 1896 by the University of Pennsylvania for the scientific study and remedial treatment of defects of development.

"During the early years of its existence the psychological clinic was open for a few hours on one day of each week. As the knowledge of its work grew the demand increased, and soon the clinic was open for three days of each week. Although the experiment of holding a daily clinic was first tried in the summer of 1897, during the six weeks of the Summer School, it was not until the fall of 1909 that regular daily clinics were established. About three new cases a day are seen. The number which can receive attention is necessarily limited, owing to the fact that the study of a case requires much time, and if the case is to be properly treated, the home conditions must be looked into, and one or more social workers employed to follow up the case. The progress of some children has been followed for a term of years."

Besides the clinic, the University of Pennsylvania also maintains a hospital school for retarded children, accepting patients as in any hospital, and giving them the psychological and physical treatment necessary to overcome their intellectual defects. Similar work has been undertaken elsewhere, particularly at Clark University, where President G. Stanley Hall has organized a " children's institute " for the scientific investigation of the development of school children; at the University of Washington, where a psychological clinic was started in the winter of 1909-10 under the direction of Professors H. C. Stevens and E. O. Sisson; and in the city of Los Angeles, which has established a department of health and development, with a director in charge of a psychological clinic. It is an educational innovation that is destined to be widely adopted. The investigations of Professor Witmer and his assistants, as well as of other psychologists and educators, have shown that there is an amazing degree of mental retardation among the school-going population of the United States. Thus, one investigator, Dr. Oliver P. Cornman, Associate Superintendent of the Public Schools of Philadelphia, has found that in five representative American cities from 21.6 per cent to 49.6 per cent of the elementary school population are one year and more behind the grade in which their age should have placed them, that from 7.3 per cent to 26.3 per cent are two years and more behind grade, that from 2.1 per cent to 12.7 per cent are three years and more behind grade, and that in one city as high as 5.1 per cent are four years and more behind the grade in which they ought to be.[1]

---

[1] Oliver P. Cornman's "The Retardation of the Pupils of Five City School Systems," in *The Psychological Clinic*, vol. i.

Not all of these backward children are susceptible of improvement, for sometimes their deficiencies represent a congenital feeble-mindedness which not even the most skillful educational methods can remedy. But in the great majority of cases, as the results obtained in Professor Witmer's psychological clinic and hospital school indicate unmistakably, the trouble is due to remedial causes. The teacher may be at fault, or, as often happens, the child may be suffering from some physical trouble, in itself slight, but sufficient to affect his mental development adversely. Eye, throat, nose, ear, and dental trouble, it has been conclusively demonstrated, are frequently productive of marked intellectual deficiency.

A typical case in point is that of a small boy who was brought to Professor Witmer's clinic two years ago with a lamentable history of intellectual backwardness and moral obliquity. Psychological examination satisfied Professor Witmer that the boy was neither a mental nor a moral imbecile, as had been suspected, and at first nothing abnormal was found in his physical condition. But it was later discovered that he was suffering from dental impaction, and it was deemed well to remove a few of his teeth. Remarkable improvement, both mental and moral, at once followed. The boy was closely observed, given some preliminary training, and then placed in a private school for education along lines laid down by the psychological clinic.

"His whole demeanor under the private instruction," says Dr. Arthur Holmes, an assistant of Professor Witmer's, who has been following the case closely, "has been that of a normal boy. He has been put upon his honor, and in every case he has justified the expectations of his teacher. He is now a healthy boy, with a boy's natural curiosity, with good manners, good temper, with no more than the average nervousness, and with every prospect of taking his proper place in society and developing into an efficient and moral citizen."

In another case two boys, twins, eight years of age, were taken to the clinic for observation. One was a bright-looking lad, sturdy, and with an excellent record at school. The other was ill-nourished, had never been to school, and looked and was supposed to be feeble-minded. But he responded to Professor Witmer's tests with an alertness and intelligence that proved that his mental faculties were unimpaired and only required development. He was given special training, and was also, as is always done at the clinic, subjected to a rigid physical examination. It was then discovered that he was slightly deaf and was suffering from adenoids, the removal of which was followed by a rapid improvement in his mental condition, thus indicating that his retardation had been largely due to the adenoids.

When the fault lies with the teaching methods employed, Professor Witmer takes in hand the instructors of the children brought to him, and explains to them what is required. In this way the psychological clinic serves another useful purpose by disseminating sound information regarding the principles of scientific pedagogy. How helpful in this respect the psychologist can be to the educator may best be appreciated by reading, for example, Prof. Hugo Münsterberg's inspiring book, " Psychology and the Teacher." As Professor Münsterberg frankly admits, psychology cannot aid in determining the purpose, the ends, of education. That is a problem for ethics. But it can and does throw a flood of light on the correct methods to be adopted in attaining whatever ends the educator has in view. Already hundreds of teachers in this country and abroad have profited from the studies of the mind of the child worked out during recent years by such eminent psychologists as Profs. J. Mark Baldwin, G. Stanley Hall, J. Sully, and W. Preyer, and from the experiments on memory, will, attention, etc., undertaken in psychological laboratories. Professor Münsterberg does not overestimate the importance of psychology to the educator when he says:

"The teacher must know what he is to teach, and must know how to teach it, and that involves his understanding the child and all the factors which come in question when the child is dealt with. Hence the true teacher needs not only an understanding of the purposes and aims of education and an enthusiastic devotion to those ideal aims, but he needs a thorough understanding of the ways in which the mind of the child can be influenced and developed. Ethics could teach him only those purposes and ideals. If the teacher seeks insight into the means by which the aim can be reached, into the facts by which the child can be molded, his way must lead from ethics to psychology."

For the parent as for the teacher, psychology has a message of the utmost importance. If only for the discovery of the far-reaching influence of "suggestion" in the affairs of daily life, and of methods whereby this influence may be utilized to promote the mental and moral development of the child, every parent is heavily in the debt of modern psychology. Psychological experiment and observation have demonstrated that every detail in one's environment — one's friends, the books one reads, the pictures one looks at, even the paper on the walls of one's house — is of suggestive value, leaving impressions upon the mind, and especially upon the plastic mind of the child, that may persist throughout, and affect the entire course of one's after life. An interesting and eloquent fragment of testimony as to this power of childhood impressions to influence later life occurs in the writings of that famous English psychical researcher, Frederic Myers.

"The first grief that I remember," says Myers, "came from the sight of a dead mole which had been crushed by a cartwheel. Deeply moved, I hurried back to my mother and asked her whether the little mole had gone to heaven. Gently and lovingly, but without doubt, she told me that the little mole had no soul and would not live again.

"To this day I remember my rush of tears at the thought of that furry, innocent creature, crushed by a danger which I fancied it too blind to see, and losing all joy forever by that unmerited stroke. The pity of it, the pity of it, and the first horror of a death without resurrection rose in my bursting heart."[1]

Here the impression left was so profound as to exercise at least a contributory influence in giving rise, in after years, to an earnest desire on Myers's part to prove that there is indeed life beyond the grave, and from this desire resulted scientific discoveries of great importance, as will be shown in a succeeding page.

As to the effect of environmental impressions, even when not consciously apprehended, much evidence might similarly be quoted. Thus Dr. Louis Waldstein, an authority on the "subconscious," says:

[1] F. W. H. Myers's "Fragments of Prose and Poetry."

"The refined tastes and joyous dispositions of the children in a family with whom I often came into contact was a matter of some surprise to me, as I could not account for the common trait among them by the position or special characteristics of the parents. They were in the humblest position socially, and all but poor. My first visit to their modest home furnished me with the natural solution, and gave me much food for reflection.

"The children — there were six — occupied two rooms into which the sunlight was pouring as I entered; the remaining rooms of the apartment were sunless for the greater part of the day; the color and design of the cheap wall-paper were cheerful and unobtrusive, bits of carpet, the table-cover, and the coverlets on the beds were all in harmony, and of quiet design in nearly the elementary colors. Everything in these poor rooms of poor people had been chosen with the truest judgment for æsthetic effect." [1]

---

[1] Louis Waldstein's "The Subconscious Self."

In other words, poor though they were, the parents had contrived, through neatness, good taste, and the judicious distribution and arrangement of their belongings, to give their children a material environment rich in cultural suggestions. Of course in their case this had been done instinctively, and without any aid from psychology. On the other hand, many parents are unwittingly doing grievous injury to their children through ignorantly subjecting them to harmful environmental suggestions; while others, again, though as yet all too few, are directly profiting from the discoveries of psychology by becoming acquainted with them and giving them practical application.

Manifestly, this is a field in which the social worker, equally with the educator and parent, can make use of the results of psychological research. It is therefore pleasant to be able to record that the social workers of the United States are awakening to their opportunity. In the psychological clinic of the University of Pennsylvania, as we have seen, the social worker labors hand in hand with the psychologist and pays especial attention to the environmental influences surrounding the retarded children in their home life. Prof. Francis Greenwood Peabody, head of the Department of Social Ethics in Harvard University, tells me that in the courses of instruction in the Harvard School for Social Workers emphasis is laid on the importance of psychology to the social worker, and the effort is made to equip the students of the school for making use of psychological facts in dealing with the difficult questions that will confront them when they enter upon their life's work. The same may be said of other institutions of this kind, and already such appreciable results have been obtained that it is safe to hazard the assertion that psychology will eventually go far towards solving the ever-present problem of the slums.

In other directions psychology is assisting in the noble work of social amelioration. The discovery of the therapeutic value of hypnotic and hypnoidal suggestion, and of suggestion skillfully applied in the waking state has provided society with a wonderful agency not only for combating the spread of mental and nervous disease but for rescuing the victims of drug and alcoholic excesses, and of overcoming temperamental defects leading to immorality, vice, and crime. Besides which, psychological experiments have provided social reformers with new and most persuasive arguments in their campaigns against existing evils. Thus, in a noteworthy address to the Massachusetts No-License League, Pres. Charles W. Eliot took his stand squarely on the findings of psychological investigation.

"It is well known," said he, "that alcohol, even if moderately used, does not quicken the action of the mind or enable one to support mental labor. We have had a great deal of German investigation and some American investigation in psychological laboratories in that direction, and the results are perfectly plain, and they are all one. The effect of alcohol on the time reaction of the human being has been studied carefully, tested in hundreds of thousands of cases, and there is no question about the ill-effect of alcohol even in very moderate doses on the time reaction. That means that alcohol in moderate doses diminishes the efficiency of the workingman in most instances and makes him incapable of doing his best in the work of the day."

In the experiments to which President Eliot referred the subjects were first tested for their mental and physical alertness before drinking any intoxicating liquors, the time of their reaction to different stimuli being carefully measured by the chronoscope. They were then given varying amounts of intoxicants and again tested, with the result that the chronoscope revealed a distinct diminution in the rapidity of their reaction time.

The principle involved in these experiments has been applied in other ways, and promises to be extremely useful, particularly, perhaps, through utilization of the so-called "association reaction method of mental diagnosis," which, although given practical trial for the first time less than six years ago, has been strongly indorsed as valuable for several purposes.

The association reaction method is based on the theory that disquieting ideas in a person's mind will reveal themselves by variations in his reaction time and in the nature of his responses, if, for instance, he is given a list of carefully selected words and is asked to utter, after hearing each, the first word that happens to come into his head. To test the validity of this theory many experiments have been tried in European and American psychological laboratories, and the experimenters have been greatly impressed with the detective value of the method. Some of them, in fact, have made use of it in other than a merely experimental way, and with equal success. On at least two occasions the scientist who first employed it for general purposes of psychological investigation, Dr. Carl G. Jung, the distinguished neurologist of Zurich, used it to good effect to trap a thief.

One of Dr. Jung's patients had confided to him his fear that he was being systematically robbed of small sums of money by his nephew, a young fellow of eighteen. It was arranged that the young man should be sent to Dr. Jung, ostensibly to undergo a medical examination. On his arrival he was told that in order to test his mental state he was to respond, as quickly as possible, to a list of one hundred words, which Dr. Jung read to him one by one. Most of these words were quite trivial, but scattered among them were thirty-seven which had to do with the thefts, the room from which the money had been taken, or possible motives for robbery. As measured by the chronoscope, the differences in his reaction time to the harmless and to the significant words were startling.

Dr. Jung said "head," he responded — or, to put it technically, associated — "nose;"

Dr. Jung said "green," he associated "blue;" Dr. Jung said "water," he associated "air;" and so on, the average reaction time being 1.6 seconds. But it took him 4.6 seconds to find a word to associate with "thief," 4.2 seconds for an association with "jail," and 3.6 seconds for one with "police." In other cases there was an abnormally quick reaction to significant words, followed immediately by a tell-tale slowing up in the reaction to the next two or three trivial ones. When he had gone through the list, Dr. Jung sternly told the young man that he found his health excellent but his morals bad, accused him of stealing from his uncle, and, basing his assertion on the character of the reaction words, taxed him with having dissipated the proceeds of his thefts in extravagant purchases, such as a gold watch. The young man, dismayed at the seemingly supernatural knowledge of his doings displayed by the physician, broke down and made a complete confession.

By the same method Dr. Jung detected,

from among several nurses in the Zurich hospital for the insane, the one guilty of stealing a small sum of money from another nurse, and similarly secured a confession. In this case, moreover, the method had the further advantage of completely clearing from suspicion a nurse to whose guilt circumstances strongly pointed, and who would otherwise have been arrested and accused of the theft.

On the strength of these and similar achievements it has been proposed that the association reaction method should be adopted by the courts as an aid in ascertaining the guilt or innocence of an accused person, but as yet this proposal has not been favorably received. Nor do the courts seem in any degree inclined to employ the services of psychological experts in legal proceedings, though a few conspicuous exceptions are to be noted. Everybody is aware of the rôle played by Professor Münsterberg in subjecting the self-confessed multi-murderer Harry Orchard to psychological examination; and more recently Prof. H. C. Stevens was permitted to take the witness-stand and testify as a psychological expert in a murder trial at Tacoma. But there is plenty of evidence to show that our judges and lawyers, even if disinclined to recognize the psychological expert's claims to court-room practice, are perfectly ready to avail themselves, in an unofficial way, of whatever help psychology can give them. In the campaign now in progress for the reform of American criminal law and criminal procedure, a campaign set on foot two years ago with the founding of the American Institute of Criminal Law and Criminology, some of the leading psychologists of the United States are aiding in the work of mapping out the reforms that should be brought about; while in actual legal practice individual lawyers all over the country are frequently resorting to psychologists for advice that will assist them in best serving the interests of their clients.

So far as the association reaction method

of mental diagnosis is concerned, there are many uses to which it may be put other than the detection of crime. Every physician has had the experience of being consulted by patients suffering from ailments that have their origin in secret vices which the patient is ashamed to reveal. In such a case the physician can — and some physicians do — utilize the association reaction method to get at the truth without arousing in the patient's mind the least suspicion that he is making an unconscious "confession." So, likewise, the educator and parent, armed not with a complicated chronoscope but merely with an ordinary stop-watch, can apply the method to study the mind of the child, perchance making thereby discoveries of vital importance to the little one's welfare. In this connection a story told by Professor Münsterberg in his book "On the Witness-Stand" may well be quoted. A young girl, anæmic and neurasthenic, and unable to concentrate her attention on her studies, had been sent to him for psychological advice.

"I asked her," says Professor Münsterberg, "many questions as to her habits of life. Among other things she assured me that she took wholesome and plentiful meals and was not allowed to buy sweets. Then I began some psychological experiments, and among other tests I started, at first rather aimlessly, with trivial associations. Her average association time was slow, nearly two seconds. Very soon the word 'money' brought the answer 'candy,' and it came with the quickness of 1.4 seconds. There was nothing remarkable in this. But the next word, 'apron,' harmless in itself, was six seconds in finding its association, and furthermore, the association which resulted was 'apron'—'chocolate.' Both the retardation and the inappropriateness of this indicated that the foregoing pair had left an emotional shock, and the choice of the word 'chocolate' showed that the disturbance resulted from the intrusion of the word 'candy.' The word 'apron' had evidently no power at all compared with those associations which were produced by the counter-emotion.

"I took this as a clue, and after twenty indifferent words which slowly restored her calmness of mind, I returned to the problem of sweets. Of course she was now warned, and was evidently on the lookout. The result was that when I threw in the word 'candy' again she needed 4.5 seconds, and the outcome was the naïve association 'never.' This 'never' was the first association that was neither substantive nor adjective. All the words before had evidently meant for her simply objects; but 'candy' seemed to appeal to her as a hint, a question, a reproach which she wanted to repudiate. She was clearly not aware that this mental change from a descriptive to a replying attitude was very suspicious; she must even have felt quite satisfied with her reply, for the next associations were short and to the point.

"After a while I began on the same line again. The unsuspicious word 'box' brought quickly the equally unsuspicious 'white'; and yet I knew at once that it was a candy box, for the next word, 'pound,' brought the association 'two,' and the following, 'book,' after several seconds the unfit association 'sweet.' She was again not aware that she had betrayed the path of her imagination. In the course of three hundred associations I varied the subject repeatedly, and she remained to the end unconscious that she had given me all the information needed. Her surprise seemed still greater than her feeling of shame when I told her that she skipped her luncheons daily, and had hardly any regular meals, but consumed every day several pounds of candy. With tears she made finally a full 'confession.' She had kept her injudicious diet a secret, as she had promised her parents not to spend any money for chocolate. The right diagnosis led me to make the right suggestions and after a few weeks her health and strength were restored."

Just as psychology, within the short span

of its existence as an experimental science, has proved itself eminently serviceable to the parent, the educator, the social reformer, the judge, and the lawyer, so has it also demonstrated its helpfulness to the business man. Take, for instance, the case of the merchant, the man who has goods to sell. In order to sell them he must bring them to the notice of the public, and in order to do that he advertises. An immense amount of money is annually wasted by advertisers who might have made a successful campaign had they utilized the results of the systematic investigations into the psychology of advertising conducted during the past few years in several American psychological laboratories, notably the Northwestern University laboratory, the director of which, Prof. W. D. Scott, has written two books on advertising — "The Theory of Advertising" and "The Psychology of Advertising" — that ought to be carefully studied by all advertisers.

An advertisement obviously is an appeal

to the minds of its readers. Many advertisers seem to think that the appeal is bound to be successful if only they advertise often enough. There is a sound psychological law underlying this idea, for repetition undoubtedly tends to establish an unconscious thought habit. On the other hand, psychological investigation has shown that unless great care is exercised in the wording or illustrating of an advertisement its repetition may induce a thought habit wholly unfavorable to the article advertised. Not only the wording, the illustrating, the position, but even the kind of type used and the general typographical appearance may be decisive of success or failure. Advertisers of course have always recognized this to a greater or less extent, but usually the process of ascertaining just what kind of advertisements they ought to adopt has been a costly one to them. They can save — and many of them to-day are saving — a great deal of needless expenditure by drawing on the expert knowledge of the psychologist, who is able, by a few experiments, to determine with a high degree of exactitude the probable effectiveness of any given advertisement. He can help the merchant, further, with respect to that special form of advertising known as window-dressing, and also with respect to salesmanship. To such an extent is this true that the day seems bound to come when every great commercial establishment will maintain a psychological laboratory of its own.

The manufacturer, the miner, the operator of transportation facilities, can likewise learn from the psychologist how to conduct their enterprises to better advantage to themselves, to their employees, and to the public. And they have begun to learn. To-day, for instance, no railway or steamship company would employ an engineer or pilot without first testing him for color-blindness, thus making practical application of the important psychological discovery of the variations of the color sense in men. But even here, the modern psychologist insists, the transportation companies do not go far enough. It is no less important, he says, that the man on the engine or at the wheel should be tested as to the rapidity of his reactions, the accuracy of his perceptions, the quickness of his decisions; and the psychological laboratory of to-day provides the instruments for making just such tests.

Only recently a Harvard psychologist, Mr. Charles Sherwood Ricker, invented a delicate apparatus for testing with the greatest precision the qualifications of the would-be automobile chauffeur. It is Mr. Ricker's contention, amply justified by facts of everyday occurrence, that hundreds of men are driving automobiles who should never be permitted to occupy the driver's seat. As he points out:

"Wherever activity on the part of one or more individuals involves questions of public safety, not only self-examination, which is ordinarily termed introspection, should be brought into play, but a psycho-physical examination is also of great importance.

Wherever quick thinking and quick acting may become a matter of life and death the discriminating hand of science should eliminate incompetent and irresponsible individuals."

Mr. Ricker designs, therefore, to aid in bringing about this desirable result by establishing a standard reaction time, which must be reached by all candidates for a chauffeur's license. His apparatus involves the flashing of certain signal lights before the candidate's eyes, the rapidity with which he responds to each signal being registered on a revolving cylinder covered with smoked paper. It is too soon to pass judgment on the availability of his invention for the purpose Mr. Ricker has in view, but the mere fact that he has invented it testifies eloquently to the earnestness with which the modern psychologist is laboring for the public good.

The same is to be said of a curious series of experiments in progress in one of the rooms of the Harvard psychological laboratory throughout the college year 1909-10. Three and four times a week members of the psychological department came to this room, two at a time, to engage for an hour or more in what would seem to the uninitiated to be merely a childish guessing game. On the table at which each couple seated themselves was a simple piece of apparatus consisting of a broad base with two uprights midway at both ends, supporting a thick cardboard top so adjusted that at a light touch it would fall to either side, preventing any view of what was taking place on the other side of the table.

While it was thus in position one of the experimenters arranged on the base six picture post-cards of different designs, and then drew the top towards him, exposing the cards to his companion, who glanced at them for a period of from three to five seconds. His view was once more cut off, the six cards were shifted about and one was withdrawn, another quite similar except for a few minor details being substituted in its place. He was now allowed to look at the cards once more, and was asked to tell which was the new card in the set. This process was repeated fifteen or twenty times, the other experimenter meanwhile making a written record of the correctness or error of his judgments, the time he took in reaching them, etc. Occasionally sets of words, printed on separate cards, were used instead of picture post-cards.

On the surface all this appears to be a waste of time, and utterly futile. In reality it is another illustration of the way psychology is being adapted to serve the needs and solve the problems of everyday life. As is well known, whenever a commodity of any sort — a food product, a beverage, whatever it may be — finds favor with the public, the market is soon flooded with imitations so put up and labeled as to deceive many purchasers into thinking that they are getting the article they really want. Sometimes the attempt to counterfeit is so obvious that legal redress may readily be had, but more often the unscrupulous imitators so word or print their labels as to raise a doubt whether they can be successfully prosecuted, the presumption being that people ought to be able, by using their eyes, to detect at a glance the difference between the article offered to them and the one that they set out to purchase.

It is chiefly to determine this point that the Harvard experiments were undertaken, and they have clearly demonstrated that far more rigid laws than now exist against commercial imitation are necessary for the proper protection of the public.

"The men with whom I experimented," says Mr. F. W. Foote, who had charge of this investigation, "had had much practice in observational work. Every one of them, when he sat down at the table with me, knew that I was going to fool him if possible, and he was on the alert to notice the least difference in the word-cards and post-cards presented to him.

"Yet I had to record an astonishingly high percentage of failures to detect substitution, amounting in some instances to more than sixty-five per cent. And this sometimes when I substituted not simply a card differing in detail, but a card of entirely different design. If trained observers can be thus deceived, it manifestly is unreasonable to expect the great untrained majority to show higher powers of discernment."

Such are some of the achievements and possibilities of the new science of applied psychology. Surely it does not need a prophet to foresee that it has a wonderful future before it.

VII

Half a Century of Psychical Research

PSYCHICAL research, as everybody knows, has for years been vainly endeavoring to gain recognition as a legitimate branch of science. Carried on largely by men of scientific temperament and training, the scientific world has nevertheless steadfastly regarded it with disfavor and suspicion, if not with openly voiced contempt, while the general public has maintained much the same attitude. There has been evident, it is true, an increasing popular interest, particularly in the United States, as not long ago was unmistakably indicated in connection with the arrival from Italy of the much talked about "medium" Eusapia Paladino; but, in the main, it is the interest of mere curiosity and love of the sensational.

Yet there are many reasons why psychical research should not only be given a friendly hearing, but should be generously supported both by scientific and by public opinion. The great Gladstone once referred to it as " the most important work which is being done in the world — by far the most important " — and this statement is by no means as extravagant as it sounds. Psychical research is not, as most people seem to take for granted, a matter merely of collecting ghost stories, watching articles of furniture move about mysteriously, and otherwise dabbling in the occult. Its great object is, of course, to determine scientifically whether or no there is life beyond the grave. But it also has another side, and one that involves considerations of the utmost practical importance to humanity.

In their quest for scientific proof of life after death the psychical researchers have from time to time made discoveries throwing a flood of entirely new light on the nature of man, and more particularly on the processes and powers of the human mind. They have confirmed, and to no small extent anticipated, the discoveries of the psychopathologists, or medical psychologists, whose activities have already been described. To them, scarcely less than to the psychopathologists, is due the present general recognition of the tremendous influence exercised by suggestion in the lives of men, and the part played by the subconscious in determining the physical, intellectual, and moral development of the individual. They have thus placed modern medicine, psychology, sociology, criminology, and pedagogy heavily in their debt.

Sometimes the debt is frankly acknowledged, and by scientists who have little or no sympathy with the ultimate aim of psychical research, believing it to be quite impossible to obtain scientific proof of the future life. Only recently our two leading psychopathologists — Dr. Morton Prince and Dr. Boris Sidis — told me that the inspiration for their life-work came to them, not from the French pioneers of psychopathological investigation, but from two English psychical researchers.

"It was through reading Edmund Gurney's reports on his experiments with hypnotism," said Dr. Prince, "that my attention was first called to the importance of studying subconscious mental states as a factor in the causation and cure of disease and that I was led to investigate them for myself."

And Dr. Sidis testified:

"My interest in psychopathology dates from the day I became acquainted with the results of Frederic Myers's preliminary studies of the subconscious. It was Myers who first opened my eyes to the close relationship between psychology and medicine."

These statements should help one to appreciate the force of Gladstone's hearty, unqualified commendation. They should also assist to an understanding of how it comes that, although psychical research has failed to win the slightest recognition from the world of science, individual scientists of the distinction of the late Professor James, Sir Oliver Lodge, Sir William Crookes, Lord Rayleigh, Professor Morselli, and the late Professor Lombroso have been found willing to engage in the investigation of its debatable phenomena.

The field covered by psychical research is indeed a broad one. When first set on foot, however, about the middle of last century, following the advent of modern spiritism as expounded by the Fox sisters and their numerous rivals and imitators, its scope was decidedly limited, being practically confined to inquiry into the so-called " physical phenomena " — the raps, table-tipping, materializations, levitations, etc. — of the séance room of that day. Nor was it then conducted on any systematic and soundly scientific basis. As a general rule, the researchers were more than half-way spiritistically inclined before they began their investigations, and, lacking knowledge of the multifarious means by which the phenomena could be fraudulently produced, were easily imposed upon, became avowed spiritists, and ceased to investigate. Or, if not overcredulous, they had their interest so thoroughly chilled by the proofs of fraud evident on every hand that they soon abandoned their inquiries as wholly useless and unprofitable.

This was the case both in the United States and in England, the two countries where spiritism first obtained a foothold. At only one place — the University of Cambridge — was sustained effort made to get at the " bottom facts." There, in 1850, under the leadership of Edward White Benson, afterwards Archbishop of Canterbury, a number of serious-minded undergraduates organized a " Ghost Society," for the investigation, not only of mediumistic, but also of all phenomena of a seemingly supernatural character — apparitions, hauntings, and the like. The members industriously attended spiritistic séances, and, by means of a widely distributed circular requesting their friends and acquaintances to report any experiences of their own that seemed at all ghostly, managed to secure a large amount of psychical data of the most varied kind. In this way the Ghost Society (although during the ten or more years of its existence it contributed nothing towards the definite settlement of the problems raised by the experiences reported to it) really did much to lay the foundations of the English Society for Psychical Research, the parent body of modern psychical research associations, and the one to which science owes most.

Indeed, there is a direct connection between the Ghost Society and the Society for Psychical Research in the fact that the famous English philosopher Henry Sidgwick, who was the first President of the Society for Psychical Research and the controlling influence in its policies from its organization, in 1882, until his death in 1900, was also the most prominent and active member of the Ghost Society during its later years. Sidgwick was less than twenty years old when he joined the Ghost Society, but he was even then, as his disciple and friend Frederic Myers has pointed out, eager to ascertain "whether the study of Oriental languages, of ancient philosophies, of history, of science, would throw light upon that traditional Revelation which hung before him with so much of attractiveness in its promises." To Sidgwick's restlessly inquiring mind the phenomena of spiritism offered another and necessary field for investigation, and he plunged into it with ardor, balanced, however, by the keenness of insight, the sanity of judgment, and the profound yet open-minded skepticism that were always his distinctive characteristics. His published correspondence of this period shows that he desired, above all things, to obtain empirical proof to buttress his wavering religious faith, and that he both hoped and doubted that in psychical phenomena such proof might be obtained.

"Spiritualism," he wrote to his friend H. G. Dakyns, "progresses but slowly; I am not quite in the same phase, as I — fancy I!

— have actually heard the raps, so that your 'dreaming awake' theory will require a further development. However, I have no kind of evidence to come before a jury. . . . I can only assure you that an evening with 'spirits' is as fascinating to me as any novel. I talk with Arabs, Hindus, Spaniards, Counts Cavour, etc. I yield to the belief at the time, and recover my philosophical skepticism next morning." [1]

Time after time he detected fraud, particularly in the production of the physical phenomena, but continued his investigations, believing that " where there is so much smoke there must be flame." Yet, the Ghost Society having been disbanded, there would, in all likelihood, have come a day when he would have grown weary of his seemingly barren labors had it not been for the chance occurrence of a visit paid to him by Frederic Myers in the winter of 1869–70.

Although still in his early manhood, Myers had already given brilliant promise

[1] "Henry Sidgwick." A Memoir by A. S. and E. M. S.

of the poetic genius, the philosophic penetration, and the marvelous command of language that eventually made him one of the greatest masters of English prose that the nineteenth century produced. But he did not come to Professor Sidgwick to discuss literary questions. He was passing through the crisis of religious doubt and anxiety so common to young men of intellect, and in the hour of his sore need he turned to his former instructor, as to one in whom, while a student at Cambridge, he had found an unfailing counselor and friend. Out of the intercourse they now had together modern psychical research of the systematic, far-reaching, scientific type came into being. Myers himself, in the beautiful tribute which, on the eve of his own death, he paid to the memory of Sidgwick, has left a glowing account of its origin.

"In a star-light walk which I shall not forget," said he, "I asked him, almost with trembling, whether he thought that when Tradition, Intuition, Metaphysic, had failed to solve the riddle of the Universe, there was still a chance that from any actual observable phenomena — ghosts, spirits, whatsoever there might be — some valid knowledge might be drawn as to a World Unseen. Already, it seemed, he had thought that this was possible; steadily, though in no sanguine fashion, he indicated some last grounds of hope; and from that night onwards I resolved to pursue this quest, if it might be, at his side." [1]

Earnestly the two friends went to work, subjecting to rigid analysis the evidence collected by the old Ghost Society, hunting out and investigating mediums, and gathering additional data from every possible source. Soon, to their satisfaction, they drew about them a little group of Cambridge men — including Edmund Gurney and Arthur J. Balfour, the latter of whom, of course, is far better known to-day as leader of the British Conservative party than as a psychical re-

[1] F. W. H. Myers's "In Memory of Henry Sidgwick," an address delivered before the Society for Psychical Research, and published in its *Proceedings*, vol. xv.

searcher — eager, like themselves, to solve, if possible, the " riddle of the Universe." And, although continually encountering fraud, the further they pressed their inquiry, the more they became persuaded that at least some psychical phenomena were not explicable by the ordinary hypotheses of deception, delusion, or chance coincidence. But how to determine their true explanation was a problem that seemed beyond them. " Our methods, our canons," as Myers afterwards explained, " were all to make. In those early days we were more devoid of precedents, of guidance, even of criticism that went beyond mere expressions of contempt, than is now readily conceived."

A serious interest in matters psychical was meanwhile becoming evident in other quarters. In 1869 the London Dialectical Society undertook an inquiry into the claims of spiritism, and two years later the famous physicist, Sir William Crookes, held a series of experimental séances with Daniel Dunglas Home, the only medium of the " physical phenomena" type who enjoys the distinction of having at no time been detected in fraud when subjected to scientific scrutiny.[1] Sir William's public declaration that he believed he had, through Home, succeeded in demonstrating the existence of a hitherto unknown force, brought upon him a storm of ridicule and contempt, but also served to give fresh heart to those who, like Sidgwick and Myers, Gurney and Balfour, felt that the time had come for an exhaustive and well-organized investigation.

Their hopes were raised still higher when, in 1876, Prof. W. F. Barrett, of the Royal College of Science, Dublin, at a meeting of the British Association for the Advancement of Science, read a paper on "Some Phenomena Associated with Abnormal Conditions of Mind," in which he described various experiments tending to prove what the early mesmerists had called "community of sen-

---

[1] A detailed account of Home's mediumship is given in the present writer's "Historic Ghosts and Ghost Hunters." See also Frank Podmore's "Modern Spiritualism," especially for references to the literature on Home.

sation " and " clairvoyance," and urged the appointment of a committee of scientific men for the systematic investigation of the phenomena of hypnotism and spiritism.

Professor Barrett's appeal fell on deaf ears as far as the British Association was concerned; but it had the effect of bringing him into close touch with the Cambridge group, and of revealing a strong undercurrent of individual opinion in favor of some such investigation as he proposed. Finally, when it had become only too apparent that no already organized scientific body would act, Barrett, Sidgwick, Myers, and Gurney took the initiative, and, at a meeting held in London in 1882, founded the Society for Psychical Research.

No better idea of its purposes and of the spirit in which it approached its task can be given than by quoting briefly from the preliminary announcement issued by its founders:

It has been widely felt that the present is an opportune time for making an organized and systematic attempt to investigate that large group of debatable phenomena designated by such terms as mesmeric, psychical, and spiritistic.

From the recorded testimony of many competent witnesses, past and present, including observations recently made by scientific men of eminence in various countries, there appears to be, amidst much illusion and deception, an important body of remarkable phenomena, which are *prima facie* inexplicable on any generally recognized hypothesis, and which, if incontestably established, would be of the highest possible value.

The aim of the Society will be to approach these various problems without prejudice or prepossession of any kind, and in the same spirit of exact and unimpassioned inquiry which has enabled Science to solve so many problems, once not less obscure nor less hotly debated. The founders of this Society fully recognize the exceptional difficulties which surround this branch of research; but they nevertheless hope that by patient and systematic effort some results of permanent value may be attained.

Five distinct subjects of inquiry were named in the announcement:

1. An examination of the nature and extent of any influence which may be exerted by one mind upon another, apart from any generally recognized mode of perception.

2. The study of hypnotism, and the forms of so-called mesmeric trance, with its alleged insen-

sibility to pain; clairvoyance and other allied phenomena.

3. A critical revision of Reichenbach's researches with certain organizations called "sensitives," and an inquiry whether such organizations possess any power of perception beyond a highly exalted sensibility of the recognized sensory organs.

4. A careful investigation of any reports, resting on strong testimony, regarding apparitions at the moment of death, or otherwise, or regarding disturbances in houses reputed to be haunted.

5. An inquiry into the various physical phenomena commonly called spiritualistic, with an attempt to discover their causes and general laws.

Of the five groups of phenomena thus designated for investigation, inquiry is still proceeding with respect to the first, fourth, and fifth, and, in part, the second. The investigation of the so-called "Reichenbach sensitives" has long been abandoned, it being pretty definitely determined that the phenomena reported were due chiefly to "unconscious suggestion." Of late years, too, the Society has ceased to study hypnotism, or, to be more exact, has delegated that task to its medical members, precisely because, as was said above, medical men, largely in consequence of the early experiments conducted by the Society for Psychical Research, have recognized that the investigation of hypnotism should be carried on by them, and that in it they possess a therapeutic agent of great usefulness.

To be sure, it was not primarily from any desire to demonstrate the therapeutic value of hypnotism that its investigation was undertaken by the Society for Psychical Research. It was rather because the hope was felt that through the study of hypnotism light might be thrown on the first problem marked for solution — namely, the possibility that thought might be communicated from mind to mind without passing through the ordinary recognized channels of communication. The evidence already assembled by the little Cambridge coterie, including, as it did, hundreds of well-authenticated instances of apparitions and other coincidental visions beheld in dream or waking hallucination at the time of the death of the person seen, seemed to point strongly in this direction; while the records of the early French and German, English and American mesmerists, with their weird but constantly recurring tales of clairvoyance, indicated that, if there really were such a thing as "thought transference," a somnambulic state was a favoring condition to it. Accordingly, as soon as possible after the founding of the Society, three special committees were set to work, one exploring the mysteries of the hypnotic trance, the second attempting experiments in thought transference in the waking state, and the third sifting the evidence for spontaneous thought transference, as in apparitions, coincidental dreams, etc.

The detailed records of the experiments may be read in the Society's *Proceedings* for 1883-84, while the evidence for spontaneous thought transference as brought to the attention of the special committee intrusted with its examination is contained in two substantial volumes entitled "Phantasms of the Living," and written by Edmund Gurney, with the assistance of Mr. Myers and Mr. Frank Podmore. Here it need only be said that, in 1884, after carefully considering the reports of the three special committees, the Society's Literary Committee felt justified in affirming that " our Society claims to have proved the reality of thought transference — of the transmission of thoughts, feelings, and images from one mind to another by no recognized channel of sense." And, later in the same year, in a report dealing with the nature of apparitions, the same committee again explicitly committed the Society to belief in thought transference — or telepathy, to use the more modern term — by saying:

"Our aim is to trace the connection between the most trivial phenomena of thought transference, or confused inklings of disaster, and the full-blown 'apparition' of popular belief. And, once on the track, we find group after group of transitional experiences illustrating the degrees by which a stimulus, falling or fallen from afar upon some obscure subconscious region of the percipient's mind, may seem to disengage itself from his subjectivity and emerge into the waking world."

In this passage — written, as was said, in 1884 — we come upon a first reference to the "subconscious," that battle-ground of present-day psychological debate and controversy. In their efforts to obtain proof of telepathy the investigators of the Society for Psychical Research had unexpectedly opened up another and most difficult problem. At every turn — in the phenomena of hypnotism, the dreams, the apparitions, that they studied — they found themselves stumbling upon indications that the human mind was a far more intricate affair than was generally supposed. Nay, even the personality, the self, the ego, seemed strangely complicated and unstable, instead of being the simple, indivisible entity of philosophic and popular belief. How account for the amazing alterations it underwent in the hypnotic trance, when, at the hypnotist's bidding, the thoughts, the emotions, the memories of the waking state would be blotted out and an entirely alien personality be created? How account for the similar transformation in cases of hysteria, when one, two, or more personalities might in turn suppress and crowd out the normal self? How, again, explain the marvelous quickening and extension of mental faculty in trance, dream, reverie, when memories long forgotten in the waking state could be recalled with ease, and even scenes and events that at the time of their occurrence had not been consciously observed could be visualized and described in graphic detail? Whence, still further, came the power to apprehend, on rare occasions, yet indubitably, the unuttered thoughts of others, and to gain knowledge of happenings at a distance, as though some mental, unseen telegraph flashed the news through empty air? What did all this signify?

Various answers have been returned by

psychical researchers, by psychologists, by psychopathologists, since first this problem was raised a quarter of a century ago. But the answer that has excited the strongest interest and the liveliest discussion is that returned by the man who did most to raise the problem and devoted the remainder of his life to its solution — Frederic Myers. Few as yet accept his answer in its entirety, as given to the world eight years ago in his posthumous work, " Human Personality and its Survival of Bodily Death," wherein is formulated his brilliant, almost dazzling, conception of the " subliminal self."

Surveying the whole wide range of mental phenomena, the singular alterations and disintegrations of personality in disease, its evident limitations of faculty, counterbalanced at times by seemingly supernatural extension of faculty, Myers saw valid reason for asserting that the self of which we are normally aware — the self which one has in mind when he speaks of " my self " — is in reality only a split-off from a larger self, just as the "secondary personalities" of hypnotism and hysteria are split-offs from the self of everyday life. And to this larger self, the subliminal self, he referred, on the one hand, the intellectual uprushes and outpourings of genius and the achievements of humanity in time of stress, when, as the phrase is, a man seems to be "lifted out of himself," inspired with new energy, and capable of accomplishing deeds he had never dreamed possible to him; and, on the other hand, Myers likewise attributed to the subliminal self, as a faculty peculiarly its own, the power of telepathically transmitting messages from mind to mind and receiving and retaining them until some favoring condition permitted their presentation to the ordinary consciousness.

"I do not, indeed, by using the term 'subliminal self' assume," he explained, "that there are two correlative and parallel selves existing always within each of us. Rather, I mean by the subliminal self that part of the self which is commonly subliminal; and I conceive that there may be not only co-operations between these quasi-independent trains of thought, but also upheavals and alternations of personality of many kinds, so that what was once below the surface may, for a time, or permanently, rise above it. And I conceive also that no self of which we can here have cognizance is, in reality, more than a fragment of a larger self — revealed in a fashion at once shifting and limited through an organism not so framed as to afford it full manifestation."

It is the closing sentence, with its implications elaborated in the pages of "Human Personality and its Survival of Bodily Death," that has been the greatest obstacle to scientific acceptance of the "subliminal self." Scientists to-day, or such of them as are really acquainted with the results of the explorations of the psychical researchers and the psychopathologists in the nooks and crannies of the human mind, willingly concede that there is a vast subconscious as well as a conscious mental life, and that the study of the former is quite as important as the study of the latter. But they balk at the idea of regarding the self of ordinary, workaday existence as merely a " fragment of a larger self " hampered by " an organism not so framed as to afford it full manifestation." And they balk still more at the inference that, when the trammels of the body have been shaken off, this larger self attains complete manifestation — an inference which Myers frankly declared had become to him a certainty.

For side by side with his long and patient inquiry into the nature of personality he continued to press tirelessly his inquiry into the greater question of the survival of personality after the body's death. To him, as to his associates, this was the question of supreme importance, and he and they turned to it with a livelier hope once they had established to their satisfaction the actuality of telepathy. For, said they, if spirits still in the flesh can thus communicate with one another, is it not reasonable to suppose that, if there really be survival, disincarnate spirits can similarly communicate with their friends yet on earth? If, then, we obtain, through whatever mechanism, information purporting to come from the World Beyond; if that information is of such a character as to afford evidential proof of emanating from the spirit alleged to be communicating; and if such evidential proof is demonstrably not of mundane origin, the problem is solved and survival put beyond the shadow of a doubt.

At once attention was focused again on the phenomena of spiritism. Obviously, however, as the psychical researchers now appreciated more clearly than ever before, if proof of personal identity were what was needed, there was little to be gained from investigating the purely physical phenomena. These might be caused by spirit action, they might be caused by the action of some unrecognized natural force, they might — as had so often proved to be the case — be fraudulently caused. Whatever their cause, they had no bearing on the immediate problem at issue. The spirit of Daniel Webster or Swedenborg or Napoleon might juggle tables and shake tamborines till Doomsday without thereby affording one iota of evidence that he really was Daniel Webster or Swedenborg or Napoleon. By no stretch of the imagination could the gymnastics of the darkened séance room be interpreted as proof of personal identity.

To the mediums, therefore, who dealt not in furniture-flinging but in the communicating of "spirit-messages" by word of mouth or pen — the "automatic speakers" and "automatic writers" — the members of the Society for Psychical Research turned their attention. Not that they wholly ceased from investigating the "physical" mediums. The admirers of Eusapia Paladino, for instance, are painfully aware that, in 1895, a committee of the Society for Psychical Research caught her practicing the most unblushing fraud, and publicly denounced her as an imposter. But, for the reason stated, and for the additional reason that it is difficult, if not impossible, to persuade "physical" mediums to submit to conditions that will exclude all possibility of fraud, the psychical researchers of England and of the United States [1] have preferred to devote their time to investigating the phenomena of the automatists.

At first, however, it seemed as though trustworthy automatists were as hard to find as trustworthy "physical" mediums. Either they were detected obtaining by fraudulent means the information which they pretended to transmit from the world of spirits or else

[1] Things are different on the European Continent, where, as is well known, the "physical" mediums, and more particularly Eusapia Paladino, have almost monopolized the attention of the various eminent men of science — Flammarion, Lombroso, Morselli, Bottazzi, etc. — who have become interested in psychical research. The only explanation of this, it seems to me, is that the European investigators are less concerned with proving the survival of personality than with ascertaining the exact nature of the power displayed by the medium. And, after reading their voluminous reports, one cannot avoid the suspicion that they are, as a class, less cautious and critical than the psychical researchers of England and America. Though, as far as that goes, it is difficult to account for the action of the English Society in consenting a couple of years ago to a second official investigation of Eusapia Paladino, and thus breaking its long-established and excellent rule of having nothing further to do with mediums once detected in fraud.

it quickly became evident that they were simply " reading the minds " of their " sitters." Nor did the situation improve until 1885, when the now world-famous automatist, Mrs. Leonora Piper, was for the first time brought to the attention of the Society by Professor James, who, the previous year, had been active in the founding of an American Society for Psychical Research, an organization which, like the English Society, had some distinguished names on its membership roll, including such men as Phillips Brooks, Colonel Higginson, Andrew D. White, and Professors James, Newcomb, Langley, Royce, Pickering, Gray, and Jastrow.

In his report on Mrs. Piper to the English Society Professor James declared that he was " persuaded of the medium's honesty and of the genuineness of her trance," and that he believed her " to be in possession of a power as yet unexplained." Sidgwick and Myers and their fellow-researchers were inclined to be skeptical, as they well might be in view of the amount of fraud and chicanery they had unearthed. But, Professor James insisting on Mrs. Piper's unusual trance abilities, it was decided, in 1887, to send to America a special agent, Dr. Richard Hodgson, who had already proved his fitness for such a mission by conclusively exposing the fraudulent practices of the high priestess of Theosophy, Madame Blavatsky.

Dr. Hodgson's first step on arriving in the United States was to employ private detectives to spy on both Mrs. Piper and her husband, with a view to discovering whether either of them went about inquiring into the affairs of prospective sitters or received such information through the mail. But nothing suspicious developed, and Dr. Hodgson finally had to confess himself so baffled and perplexed that it was determined to invite Mrs. Piper to visit England and submit to investigation under conditions that would make fraudulent acquisition of knowledge impossible. The plan, as actually carried out, was to make her virtually a prisoner in the home of Mr. Myers. Her baggage was searched, her mail opened; yet at no time was evidence forthcoming even remotely suggesting that she obtained her trance information by normal means.

Not all of the séances she gave while in England were equally impressive, but at many of them her utterances seemed to be so strikingly evidential of the identity of the "spirit" purporting to communicate through her that many members of the Society felt that their quest was nearing its end. Others, under the leadership of Mr. Podmore, frankly expressed their conviction that, although no charge of fraud could be successfully laid against her, everything she had communicated of evidential value might be satisfactorily accounted for by the hypothesis of telepathy between living minds. Dr. Hodgson himself, who returned to the United States with Mrs. Piper, was strongly opposed to accepting the spiritistic view, as appeared from a lengthy report issued by him after four years more of unremitting investigation. But almost before this report was in print Mrs. Piper's mediumship entered into a new phase that shook his skepticism to its foundation.

Hitherto she had been "controlled" chiefly by a motley band of "spirits" who gave themselves Latin names, refused to reveal their identity, but claimed to act as intermediaries, so to speak, between the sitters and their deceased acquaintances. Now, following the death of a close friend of Dr. Hodgson's, the Latin "controls" were gradually ousted, and the friend's "spirit" began to take their place, giving such convincing proofs of his identity that Dr. Hodgson felt that the telepathic hypothesis would no longer suffice, and that he really was in communication with the man he had known so well in life. In 1898 he publicly announced his acceptance of the spiritistic hypothesis as the only one adequate to explain all the facts,[1] and two years later a

---

[1] See his "A Further Record of Certain Phenomena of Trance," in the *Proceedings of the Society for Psychical Research*, vol. xiii.

similar announcement was made by another investigator of Mrs. Piper, Prof. J. H. Hyslop, who believed that he had been brought into touch through her with dead relatives and friends.

From that day to the present a warm controversy has been in progress, both in England and in the United States, between the advocates of the telepathic and the spiritistic hypothesis as explanatory of the phenomena manifesting through Mrs. Piper and other automatic mediums who have since sprung into prominence — more especially certain Englishwomen, Mrs. Verrall, Mrs. Thompson, Mrs. Forbes, and Mrs. Holland. In this country, though, it must be said, there has been at no time the sustained and earnest interest in psychical research evident in England. The work of organized investigation has practically been left to four men — Professor Hyslop, the late Professor James, the late Dr. Hodgson, and, more recently, Mr. Hereward Carrington. The American Society for Psychical Research, founded under such auspicious circumstances in 1884, lapsed in 1889 into a mere branch of the English Society, and, although reorganized as an independent body in 1906, its work has since been carried on almost single-handed by Professor Hyslop. Believing firmly, as he does, in the superiority of the spiritistic to the telepathic hypothesis as explanatory of the phenomena in question, it is inevitable that the publications of the American Society should be dominantly spiritistic in tone. And they are so distinctly so that, as a matter of fact, criticism of the spiritistic hypothesis and advocacy of the telepathic comes not so much from members of the Society as from investigators not connected with it.

In the English Society a different situation prevails, though even there it almost seems as though the advocates of telepathy as against spiritism are constantly becoming fewer. Every year witnesses new accessions to the spiritistic camp. The latest " convert " is Sir Oliver Lodge, who, after more than twenty years of investigation, has at last proclaimed his belief that telepathy "is strained to the breaking point" when applied to explain all the phenomena of the Piper-Verrall-Thompson-Forbes-Holland group of mediums. Yet it would be doing the Society a grave injustice to infer that, because individual members affirm that satisfactory proof of spirit communication has been obtained, it, as a Society, indorses this view. On the contrary, from the beginning it has been consistently cautious in its pronouncements. Beyond accepting telepathy as proved — which, by the way, is as yet not the opinion of the scientific world, notwithstanding that the evidence to sustain it has been constantly strengthened with the passage of time — the Society for Psychical Research has reached almost no positive conclusions. And there seems to be no warrant for believing that it will now so far depart from the standards set by its founders as to, in the words of an indignant but hasty critic, " cease to be an organization for scientific inquiry and turn itself into an organization for the propagation of spiritism."

One criticism, however, may in all fairness be made. In concentrating their efforts on the study of the " evidential " phenomena of the automatic mediums, the members of the Society have of recent years unquestionably neglected the important field for investigation opened up by the researches of Myers and Gurney in the subconscious nature of man. This neglect, though, is probably only a passing phase, and one day we shall likely find them, under the inspiration of some second Myers or second Gurney, probing once more into the mysteries of the "subliminal" with results still more beneficial to mankind, and adding appreciably to the Society's present record of solid and valuable achievement.

VIII

William James — An Appreciation

WITH the death of William James there has passed from among us the greatest leader of American philosophic thought since the time of Ralph Waldo Emerson. Indeed, I am almost tempted to describe his death as the removal of the greatest of contemporary Americans. Certainly no other of his generation exercised such an international influence as did William James. Scholars in England, in France, in every European center of learning as well as in his own land held him in the highest esteem. No other American thinker — particularly during the last few years of his life, which he devoted so zealously to the propagation of the gospel of pragmatism — was so widely quoted as he; no other commanded so large and so respectful an audience; and the views of no other were so carefully examined and so exhaustively discussed.

And this because of universal and almost immediate recognition of the fact that he had sounded a new and most important note, and breathed a new spirit into philosophy. It was ever his aim to remove philosophy from the abstract and the intangible, to make it real and concrete, to give it meaning and vitality not merely to philosophers, but to the layman. In order to do this, he plainly saw, philosophy as a science must be correlated with the facts of experience; it must deal with things as they are, must interpret them lucidly, make clear their respective values, and act at once as mentor and as friend. Philosophy, in other words, must be practical, must cease voyaging through the clouds of abstraction and speculation, and come down to the solid earth of actuality.

It was in this spirit, and from this point of view, that Professor James began his attack on the orthodox systems of philosophy, and more especially on the dominant school of "logic-chopping" Hegelianism. I say "attack," but, after all, his effort was not so much to undermine as to reform and revivify. In his espousal of pragmatism — of which, though not the founder, he was easily the most influential advocate — he did not seek to establish a new philosophic system so much as a new outlook, a broader activity for philosophy. Pragmatism, as is now well known, is essentially a philosophy of action, of practicability. The pragmatist would test the truth, the meaning, the significance of things by their workability. His great question concerning every proposed generalization is, "Does it work?" If it works, if it is useful, then it is true.

On this basic principle, this testing of truth by its practical consequences, Professor James squarely took his stand, unmoved by the storm of controversy and academic abuse that has been steadily growing in volume since his first promulgation of the pragmatic doctrine in 1898. In the meantime, however, pragmatism itself has been gaining adherents, if only for the reason that there has been increasing recognition that we are every one of us at bottom pragmatists. In our daily lives, in solving the smaller and larger problems of existence, we constantly put them to the test of workability, of usefulness. Even the modern Hegelians, who have been the heartiest opponents of the pragmatic method, unconsciously adopt it, contending as they do for the supreme value to mankind of their idealistic philosophy.

One great obstacle to general acceptance of the doctrine so ably upheld by Professor James and his two best-known fellow-pragmatists — Professors Dewey and Schiller — lies in the question of standards. Critics have accused the pragmatists of including among the "consequences" that give meaning and vitality and truth to an idea only such as are practical in a material sense — " bread-and-butter consequences," as Professor Pratt has jestingly called them. But it is a chief merit of Professor James's work that he has consistently given primacy to the spiritual and the intellectual. This appears most clearly perhaps in his wonderful book, " The Varieties of Religious Experience," a volume which, if he had written nothing else, would give him a secure place in the history of philosophy.

It is, I may remind my readers, a scientific study of the phenomena of religious experience, with a view to accounting for religion and estimating its value. In this respect it is unique, and, in the words of one competent reviewer, " furnishes the most powerful antidote to the cynical and pessimistic scepticism of the age, since Martineau's ' Study of Religion,' which it equals in spiritual beauty and surpasses in wide observation and dramatic interpretation of the actual spiritual experiences of human souls." What Professor James did, in studying the significance of religion systematically, was to appraise it by the pragmatic method. Has religion " worked," has it been " useful," were leading questions he put to himself, and, basing his answer on the tangible facts of concrete human experience, he found himself impelled to reply to both questions with an emphatic affirmative. Moreover, his analysis led him to the firm belief that religion would endure. Religion, he declared in effect, unquestionably forms part of man's normal life, and since it also contributes to the preservation, to the integrity, and to the prosperity of that life, reason combines with instinct and tradition in favoring its continuance.

Showing himself in this book profoundly religious-minded, William James likewise showed himself to be open-minded to a degree not commonly displayed by philosophers. He could not, in truth, be a consistent pragmatist without being a man of most open mind, for to your true pragmatist dogmatism and prejudice are above all else to be avoided. But in the case of Professor James, temperament was superadded to make him open-minded to an exceptional extent. Thus, as everybody is aware, he regarded calmly, philosophically, and investigatingly matters which the majority of his colleagues, philosophers and psychologists alike, considered utterly beneath their notice. Professor James, with a generous and wise catholicity, saw in these same matters facts in human experience to be inquired into, tested pragmatically, and evaluated accordingly.

In this way, for instance, he was led more than twenty-five years ago to begin the labors in psychical research with which his name has been conspicuously associated in the popular mind. Many of his associates, nay, even many of his warmest personal friends, felt that in devoting the time he did to psychical investigations he was wasting precious hours which he might otherwise have employed to far greater profit. In reality, the world has been the gainer by the researches that brought upon him such a flood of hostile criticism, and that were, as I happen to know, prosecuted by him as much from a sense of duty as from personal enthusiasm and desire.

The world, I say, has been the gainer, and richly the gainer, by the psychical researches of William James. If his numerous séances with Mrs. Piper and other celebrated mediums, his repeated excursions into the tangled wildernesses of automatic writing and speaking, clairvoyance and clairaudience, and kindred phenomena, failed to bring to his receptive yet discriminating mind the evidential proof he sought of the survival of human personality after bodily death, they at least opened to him new vistas of psychological knowledge and philosophical insight which he has passed on to others both by the written and by the spoken word. It is not too much to say that had it not been for his delvings in the occult and the abnormal his masterwork, "The Principles of Psychology," would have lost much of the substance that, upon the instant of its appearance, gained for it recognition as one among the most stimulating and soundly informative of psychological text-books. Had it not been for these same delvings I am convinced that the "Varieties of Religious Experience," which personally I rate only second in importance to the "Psychology" and the "Pragmatism," could not have voiced the inspiring conclusions it reached. To say, as many do, that psychical research was simply a "hobby" of Professor James's, is to miss entirely the purpose for which he undertook it and the thoroughly practical results it brought to him and, through him, to his fellowmen.

Somebody once asked him what he expected to gain from his investigations into the phenomena of spiritism. "Balm for men's souls," was his instant reply. Like Myers, Sidgwick, and their fellow-founders of the Society for Psychical Research, it was to him a thing incredible that phenomena alleged to have a direct bearing on the problem of chiefest importance to man — the problem of the survival of human personality after the death of the body — should not be made the subject of the most searching inquiry. In this belief he made psychical research one of his main activities from 1884 to the time of his death, although forced to admit, in a magazine article written only a short time before his death:

"For twenty-five years I have been in touch with the literature of psychical research, and have had acquaintance with numerous 'researchers.' I have also spent a good many hours (though far fewer than I ought to have spent) in witnessing (or trying to witness) phenomena. Yet I am theoretically no 'further' than I was at the beginning; and I confess that at times I have been tempted to believe that the Creator has eternally intended this department of nature to remain baffling, to prompt our curiosities and hopes and suspicions, all in equal measure, so that, although ghosts and clairvoyances, and raps and messages from spirits, are always seeming to exist and can never be fully explained away, they also can never be susceptible of full corroboration.

"The peculiarity of the case is just that there are so many sources of possible deception in most of the observations that the whole lot of them may be worthless, and yet that in comparatively few cases can aught more fatal than this vague general possibility of error be pleaded against the record. Science meanwhile needs something more than bare possibilities to build upon; so your genuinely scientific inquirer — I don't mean your ignoramus 'scientist' — has to remain unsatisfied. It is hard to believe, however, that the Creator has really put any big array of phenomena into the world merely to defy and mock our scientific tendencies; so my deeper belief is that we psychical researchers have been too precipitate with our hopes, and that we must expect to mark progress not by quarter-centuries but by half-centuries or whole centuries." [1]

But if, so far as concerned the securing

[1] *The American Magazine*, vol. lxviii.

of scientifically acceptable proof of life beyond the grave, Professor James, after twenty-five years of patient investigation, had to confess himself baffled, his psychical researches were, as was said, none the less productive of important results. For one thing, they enlarged his understanding of the nature of man to an extent that would have been impossible had he shared the intellectual timidity, the "superstition of incredulity," common among men of science when the so-called occult is in question. The marvelous extensions of human faculty observable in the phenomena of the spontaneous and the induced trance; the evidence of "subconscious" powers and processes, manifest in automatic speaking and writing, in dreams, in the psycho-physiological effects of "suggestion," gave him a clearer insight into the make-up and possibilities of personality than would ever have been his had he refrained from investigation.

So likewise with the interest he took in Christian Science and the New Thought.

Pragmatically speaking, they appealed to him because they "worked." But he saw clearly enough that they did not always "work"; that they had many failures as well as many "cures" to their account; and, probing into the problem further, he was brought into direct contact with the scientific mental healing, the psychopathology of Liébeault and Bernheim, of Charcot and Janet, that has already profoundly influenced the practice of medicine. Himself a physician as well as a psychologist — his first years as a teacher at Harvard were devoted to instruction in comparative anatomy and physiology — Professor James was quick to appreciate the importance of the discoveries of the French suggestionists. Probably no other American has done as much as he in the way of disseminating information as to the exact rôle played by the mind in relation to the health and disease of the body.

All attempts, therefore, to belittle the achievements of William James by reason of his intimate association with psychical research and mental healing must signally fail when due regard is paid to the results of this association as exemplified in his life and writings; similarly with what would seem to be a growing tendency to depreciate his work as a psychologist. It is quite true that Professor James had scant sympathy with the ultra-experimental psychology now regnant wherever Germanic influences prevail. But it should be remembered that one may very well be a psychologist without surrendering all his time to the manipulation of the chronoscope and sphygmograph and allied ingenious devices for digging into the human mind. In fact, the ideal psychologist must also be a philosopher, able to perceive not only the trees, but the forest that they constitute.

Not that I would assert that the machine-manipulating psychologist has no useful rôle to perform. He has, indeed, effected a sorely needed revolution in psychological methods, as the effort was made to indicate in the essay on " Psychology and Everyday Life."

Experimental psychology, however, is by no means the whole of psychology; it involves much more, and Professor James stood *par excellence* as representative of that larger whole. Familiar with the instruments of psychological experimentation, not disdaining to use them as occasion required, he resolutely and properly refused to be trammeled by laboratory methods, and instead went freely into the open world to observe, to examine, to investigate mental phenomena as they revealed themselves in the ordinary interplay of daily, human experience.

It is absurdly wide of the mark to suggest, as more than one critic has suggested, that his psychology is but transitional, and in time will have only a historical interest. For in his work, as perhaps nowhere else, we find psychology laboring hand in hand with philosophy to explain, to interpret, to make luminously clear the phenomena of the human mind and of the human soul. It is chiefly this — his abundant recognition of the spiritual as well as the mental and physical in man — that gives and will continue to give vitality to the psychological teachings of Professor James. And this, alas, is precisely what is lacking in the teachings of many of those who mistakenly regard him as the exponent of a "transitional" psychology.

They have, unquestionably, ground not for complaint — in which some of them indulge — but for admiration and emulation, in the fact that he was the possessor of a superb literary style, a style so lucid, so simple, so attractive as to gain for him an attentive and intelligent hearing in quarters where psychology and philosophy usually make little or no impression. To speak of this style of his as a gift would be scarcely accurate, for there can be no doubt that he deliberately and sedulously cultivated it. The animating principle of his intellectual life, as has been said, was to make philosophy real and helpful to the everyday man, and he knew full well that to accomplish this it must be presented in terms understandable by the everyday man. Here, of itself, was an incentive for him to avoid the abstract, to deal always with the concrete, to stick closely to life even at the cost of sacrificing logic. "When," to quote a good story told of him by Dean Hodges, "he is tempted to follow his argument into regions where logic takes the place of life, 'I heard,' he says, 'that inward monitor of which W. R. Clifford once wrote, whispering the word "Bosh!"'" And, as Dean Hodges adds, it was his insistence on the concrete that made him the most interesting as well as the most intelligible of all our contemporary philosophers.

The concrete, " after all, is what we care for. That is what commands our middle-class attention. The abstract may be profound, it may be a necessary form of philosophical expression, it may be true, but it is a foreign language. Whoever uses it begins to speak in the Hebrew tongue. The concrete is the vernacular. Whenever we hear it in a lecture, in a sermon, in a printed book, we sit up and listen. Professor James thinks in it and speaks in it. This is a great part of the secret of the singular charm of his style, in which he unites the dialect of psychology with the idioms of common speech. 'The God whom science recognizes must be,' he says, 'a God of universal laws exclusively,' to which philosophical statement he adds an immediate translation, 'a God who does a wholesale, not a retail, business.'" [1]

In other words, Professor James as a writer was the very reverse of a dry-as-dust pedant. Nor was there anything of the pedant in him as a class-room instructor. Just a week before the sad news came from New Hampshire — news which, I am free to confess, came to me with a sense of deep personal loss — I was dining with a friend who in years gone by had studied psychology under Professor James at Harvard. The conversation touched on his methods as a teacher.

[1] George Hodges's "William James," in *The Outlook*, vol. lxxxv.

"One never got the impression," my companion observed, "that he was listening to a lecture. In fact, Professor James did not lecture. He simply took his seat, started talking about his subject in a conversational way, and pretty soon some of us were talking about it also. That was his idea, that was his plan of education. He wanted to interest us, to draw us out, and I can assure you he succeeded. He got more from me than any other instructor at Harvard did — and I know that I got more from him than from any other instructor."

It was not the first time I had listened to testimony like this, giving voice, more or less eloquently, to the tremendous inspirational influence exercised by William James as a teacher, and to the love and reverence in which his former pupils held him. From what some of them have told me, and from what I have myself observed, it is easy to understand why, at the time of his retirement from active academic life, the men in his largest class united in presenting him with a loving cup; and why, when he delivered the last of his Gifford lectures at Edinburgh University, according to one of his auditors, "the crowd cheered lustily as he mounted the dais; and he left the room amidst the hearty singing of 'He's a jolly good fellow,' started and carried through by the students who were present."

The same qualities that endeared him to his pupils — his quiet manliness, his transparent sincerity, his passionate devotion to truth, his unfailing sense of humor, his open-mindedness, his catholicity, modesty, and geniality — bound him closely to his friends of maturer years. He was always a man, a real human man, first; and a philosopher afterwards. I have often heard it said that a stranger, meeting him for the first time, and approaching him with the awe due to one of a world-wide reputation, forgot all about his reputation after five minutes' conversation with him. He had the faculty, frequent among men of the world, but rare among scholars, of meeting all comers on their own level, and making them instantly feel at ease. Yet there was that about him — a quiet dignity, a fine reserve — that forbade any undue familiarity. He was a " jolly good fellow," as the Edinburgh students sang. He was never a " hail fellow well met " in the ordinary meaning of the term.

Revealing in his books and in his lectures a vast store of erudition and a rich fund of human sympathy, he gave freely of his erudition and sympathy in private life. None who sought him for advice and assistance, whether moral, intellectual, or material, sought him in vain. He was the soul of hospitality, as many a luckless scholar, many a struggling author, can testify. When a word from him would help along a man or a book in which he believed — and he believed in many men and many books — that word was never withheld. Yet for all his charitableness to men and to ideas he was not readily deceived. Perceiving with incisive insight philosophic shams, he saw as clearly through the triflers of life, the spiritual humbugs, impostors, and ne'er-do-wells that occasionally darkened his doors. Even such, however, so large, so generous was his nature, he could not treat unkindly.

Now he is gone. Now the home among the elms of classic Cambridge will know him no more. Never again will his warm handclasp greet the scholar from abroad, the colleague from across the way, the eager, ambitious student. Never again will you or I meet him taking his afternoon stroll, head erect, eyes beaming, beard bristling. He is gone — gone to that unseen world whose mysteries he so patiently explored in the hope that mayhap from the exploration he might gain "balm for men's souls." But the memory of him as a man, a teacher, a friend, will linger with us; and long after we too have passed beyond the other side of the veil, the fruits of his life's labors will remain.

For William James, as a psychologist, as

a philosopher, was no mere meteor in the intellectual firmament. Rather, his place will be as that of a fixed star of the first magnitude.

INDEX

Aboulia, case of, 78–85.
Animal magnetism, 8.
Association reaction method, 176–84.

Baldwin, J. M., 167.
Balfour, A. J., 204.
Barrett, W. F., 206, 207.
Beauchamp, C. L., 153n.
Beaunis, H., 68.
Benson, E. C., 199.
Bernheim, H., and post-hypnotic commands, 113–4; also mentioned, 68, 122, 242.
Bertrand, A., 13, 15.
Bottazzi, Prof., 221n.
Bourne, A., case of secondary personality, 145–7.
Braid, J., investigates mesmerism, 14–5; invents term hypnotism, 16; also mentioned, 22.
Bramwell, J. M., hypnotic experiments by, 111 and n, 117–8.
Breuer, J., 94.
Brill, A. A., 100 and n.
Brooks, P., 222.
Bruce, H. A., 17n, 141n, 206n.
Buchanan, J. R., 21.
Burckmar, L., mesmeric diagnosis by, 24–5.
Burnett, S. G., case of secondary personality reported by, 127–36.

Carrington, H., 226.
Carroll, H. K., and Christian Science statistics, 32n.
Charcot, J. M., researches and discoveries in hysteria, 67–70; also mentioned, 19, 39, 41, 45, 94, 242.
Christian Science, history, 23–31; statistics, 32n; compared with New Thought, 32–3; compared with scientific psychotherapy, 46–7, 64–5; also mentioned, 19, 241.
Colquhoun, J. C., 4n.
Coriat, I. H., 90n.
Cornman, O. P., and mental retardation, 163 and n.
Crane, A. M., 36.
Crookes, Sir W., investigates D. D. Home, 205–6; also mentioned, 198.

Dakyns, H. G., 201.
Dana, C. L., case of secondary personality reported by, 141–5.
Davis, A. J., 22.
Dewey, J., 233.
Digby, Sir K., 5, 6n.
Dissociation, psychical, 41–3, 48–64, 69–100, 124–55.
Donley, J. E., case of hysteria cured by, 63–4.
Dresser, H. W., 25n, 35, 36.

Dresser, J. A., and founding of New Thought, 35-6.
Dubois, P., and psychic re-education, 91.

Ebers papyrus, 2.
Eddy, M., cured by psychotherapy, 27; theorizes on psychotherapy, 28; founds Christian Science, 29-32; also mentioned, 33. See also Christian Science, James, Psychotherapy.
Eliot, C. W., and liquor question, 174-5.
Emerson, R. W., 230.
Energy, doctrine of reserve, 86.
Evans, W. F., and founding of New Thought, 35.

Faith Healing. See Psychotherapy.
Faria, Abbé, and role of suggestion in hypnotism, 13, 15.
Flammarion, C., 221n.
Folie de doute, case of, 78-85.
Foote, F. W., and business psychology, 192-3.
Forbes, Mrs., 226, 228.
Forel, A., and post-hypnotic commands, 111-3.
Fox sisters, and modern spiritism, 22, 198.
Freud, S., researches and discoveries of, 94-101; case treated by, 95-9; also mentioned, 67, 101.
Furness, W. J., case of secondary personality reported by, 137-40.

Gerrish, F. H., and susceptibility to hypnotism, 106-7.
Ghost Society, and beginnings of psychical research, 199-202.
Gilbert, J. A., case of secondary personality reported by, 150-3.
Gladstone, W. E., and significance of psychical research, 195, 197.
Gray, Prof., 222.
Grimes, J. S., 20.
Gurney, E., helps found Society for Psychical Research, 207; co-author "Phantasms of the Living," 212; also mentioned, 204, 229.

Hall, G. S., and child psychology, 162, 167.
Helmont, J. B. v., 5, 8.
Higginson, T. W., 222.
Hodges, G., and characteristics of W. James, 246-7.
Hodgson, R., investigates Mrs. Piper, 223-6; also mentioned, 147n.
Holland, Mrs., 226, 228.
Holmes, A., 165.
Home, D. D., investigated by Sir W. Crookes, 205-6 and n.
Hypnoidization, description of, 80-2 and n; also mentioned, 50, 78, 93, 101, 158.
Hypnotism, early history, 1-3; rediscovery by Mesmer, 7-12; named by Braid, 16; investigated by Liébeault, 16-8; later investigations,

39-42, 103-23; status as therapeutic method, 104; conditions for inducing hypnotic state, 105-8; hypnotic and post-hypnotic suggestions and hallucinations, 108-18; and crime, 108-20; dangers, 119-22; applied in cases of secondary personality, 151-3 and n; studied by psychical researchers, 208-11. See also Hysteria, Personality, Psychical research, Psychopathology, Psychotherapy, Subconscious, Suggestion.

Hyslop, J. H., 226, 227.

Hysteria, causation, 41-2, 99-100; symptomatology, 43-4; Charcot's and Janet's researches, 68-70; Sidis's researches, 77-87; Prince's researches, 90-3; Freud's researches, 94-101; illustrative cases, 48-64, 70-6, 78-85, 92-3, 95-9. See also Hypnotism, Personality, Psychopathology, Psychotherapy, Subconscious, Suggestion.

Hystero-epilepsy, cases of, 48-52, 70-2.

James, W., career, personality, and contributions, 230-52; pragmatism, 231-3; psychological study of religion, 234-5; interest in psychical research, 236-41; interest in psychotherapy, 241-2; status as a psychologist, 243-5; literary qualities, 245-6; as a teacher, 246-8; personal traits, 249-51; also mentioned, 78, 147n, 198, 221-2, 226.

Janet, P., researches and discoveries of, 67-76; cases treated by, 70-5; also mentioned, 44, 67, 95, 101, 242.

Jastrow, J., 222.

Jones, E., 100.

Jung, C. G., and association reaction method, 177-9.

Kennedy, R., and Christian Science, 30, 31.

Kennon, B. R., case of secondary personality reported by, 137-40.

Langley, S. P., 222.

Lawrence, R. M., 4n.

Liébeault, A. A., investigates hypnotism, 16-9; on therapeutic value of hypnotism, 122; also mentioned, 22, 68, 242.

Liégeois, J., on possibility of hypnotic crimes, 115-6; also mentioned, 68.

Lodge, Sir O., accepts spiritistic hypothesis, 227-8; also mentioned, 198.

Lombroso, C., 198, 221n.

Martial, 3.

Maxwell, W., 5.

Mental Healing. See Psychotherapy.
Mesmer, F. A., methods of, 7-12; also mentioned, 17, 38.
Mesmerism. See Hypnotism.
Morselli, H., 198, 221n.
Münsterberg, H., and "untrue confessions," 120; and educational psychology, 167-8; and Orchard case, 179; and association reaction method, 181-4.
Myers, F. W. H., influence of childhood impressions on, 167-70; influence of H. Sidgwick on, 202-4; characteristics, 203; helps found Society for Psychical Research, 207; co-author "Phantasms of the Living," 212; doctrine of subliminal self, 215-8; and Mrs. Piper, 224; also mentioned, 111, 229, 238.

Neurasthenia, case of, 92.
Newcomb, S., 222.
New Thought, history, 23-7, 32-6; compared with Christian Science, 32-3; compared with scientific psychotherapy, 46-7, 64-5; also mentioned, 19, 241.

Paladino, E., 194, 220, 221n.
Paralysis, cases of hysterical, 63-4, 69, 76.
Patterson, C. B., and New Thought doctrine, 33-4 and n.

Peabody, F. G., and social psychology, 173.
Personality, characteristics of double, 126-7, 136, 140-1, 145, 150; cases illustrative of double, 127-53; doctrine of subliminal self, 215-7; problem of survival, 219-29. See also Hypnotism, Hysteria, Psychical research, Psychology, Psychopathology, Psychotherapy, Subconscious, Suggestion.
Phobia, cases of, 53-62, 92-3.
Pickering, Prof., 222.
Piper, L. E., endorsed by W. James, 222-3; investigated by Society for Psychical Research, 223-6; also mentioned, 228, 237.
Plautus, 3.
Podmore, F., co-author "Phantasms of the Living," 212; advocates telepathic hypothesis, 224.
Pomponatius, P., 4, 8.
Poyen, C., 20.
Pragmatism, 231-3.
Pratt, Prof., 233.
Preyer, W., 167.
Prince, M., personality and career, 87-9; researches and discoveries, 90-4; cases treated by, 92-3; also mentioned, 53, 67, 101, 109, 153n, 196, 197.
Psychic reëducation, 90-3.
Psychical research, objects, 195-6, 207-9; beginnings,

198–206; founding of Society for, 207; activities of Society for, 210–29; W. James's views on, 239–40.

Psycho-analysis, description of, 95; case treated by, 95–9.

Psychology, and medicine, 39–101, 180–84; beginnings of experimental, 157–8; and education, 159–72; and social reform, 172–5; and law, 176–80; and business, 185–93. See also Hypnotism, Hysteria, Personality, Psychical research, Psychopathology, Psychotherapy, Subconscious, Suggestion.

Psychopathology, early investigations, 39–45; illustrative cases, 48–64, 70–6, 78–85, 92–3, 95–9, 151–3. See also Hypnotism, Hysteria, Psychotherapy, Subconscious, Suggestion.

Psychotherapy, early history, 1–7; Mesmer's contribution to, 7–10; beginnings of scientific, 14–9, 39–45; beginnings of religious, 19–27; Christian Science system of, 27–32; New Thought system of, 32–6; differences between scientific and non-scientific, 46–7, 64–5; illustrative cases, 24–5, 48–64, 70–6, 78–86, 92–3, 95–9, 151–3. See also Hypnotism, Hysteria, Psychopathology, Subconscious, Suggestion.

Puységur, Marquis de, 11.

Quimby, P. P., investigates mesmerism, 23; cured by psychotherapy, 24–5; theorizes on psychotherapy, 25–7; also mentioned, 25n, 28, 34, 35.

Rayleigh, Lord, 198.
Retardation, psychological treatment of mental, 159–62; statistics of mental, 163.
Ricker, C. S., and test for chauffeurs, 188–9.
Royce, J., 222.

Schiller, F. C. S., 233.
Scott, W. D., and psychology of advertising, 185.
Secondary selves. See Personality.
Seneca, 3.
Sidgwick, H., and Ghost Society, 200; views on spiritism, 201–2; influence on F. W. H. Myers, 202–4; helps found Society for Psychical Research, 207; also mentioned, 237.
Sidis, B., career, 77–8; researches and discoveries, 78–87; cases treated by, 48–62, 78–86; also mentioned, 67, 90n, 101, 196, 197.
Sisson, F. O., 162.
Solon, 3.
Spiritism. See Psychical research.
Stevens, H. C., 162, 180.
Stevenson, R. L., 124 and n, 129.

Subconscious, the, as factor in disease, 41–5, 48–64, 70–100; environment and the, 168–72; studied by psychical researchers, 213–29. See also Hypnotism, Hysteria, Psychical research, Psychopathology, Psychotherapy, Suggestion.

Subliminal self. See Personality, Subconscious.

Substitution, method of, 75.

Suggestion, therapeutic value discovered, 15–8; phenomena of hypnotic, 40; limitations of therapeutic, 43; scientific psychotherapy by non-hypnotic, 44–5; cases illustrating application for therapeutic purposes, 48–64, 70–76, 78–85; in the home, 168–72. See also Hypnotism, Hysteria, Psychopathology, Psychotherapy, Subconscious.

Sully, J., 167.
Sunderland, L., 21 and *n*, 22.
Swedenborg, E., 220.

Telepathy, 210–3, 216, 224, 227, 228.
Temples of Health, 3.
Thompson, Mrs., 226, 228.
Trine, R. W., 36.

Verrall, Mrs., 226, 228.
Voisin, A., 106.

Waldstein, L., 170–1.
Webster, D., 220.
White, A. D., 222.
Witmer, L., describes origins and methods of the psychological clinic, 159–62; cases of mental retardation treated by, 164–7; also mentioned, 163.
Wood, H., 36.

Trieste Publishing has a massive catalogue of classic book titles. Our aim is to provide readers with the highest quality reproductions of fiction and non-fiction literature that has stood the test of time. The many thousands of books in our collection have been sourced from libraries and private collections around the world.

The titles that Trieste Publishing has chosen to be part of the collection have been scanned to simulate the original. Our readers see the books the same way that their first readers did decades or a hundred or more years ago. Books from that period are often spoiled by imperfections that did not exist in the original. Imperfections could be in the form of blurred text, photographs, or missing pages. It is highly unlikely that this would occur with one of our books. Our extensive quality control ensures that the readers of Trieste Publishing's books will be delighted with their purchase. Our staff has thoroughly reviewed every page of all the books in the collection, repairing, or if necessary, rejecting titles that are not of the highest quality. This process ensures that the reader of one of Trieste Publishing's titles receives a volume that faithfully reproduces the original, and to the maximum degree possible, gives them the experience of owning the original work.

We pride ourselves on not only creating a pathway to an extensive reservoir of books of the finest quality, but also providing value to every one of our readers. Generally, Trieste books are purchased singly - on demand, however they may also be purchased in bulk. Readers interested in bulk purchases are invited to contact us directly to enquire about our tailored bulk rates. Email: customerservice@triestepublishing.com

You May Also Like

ISBN: 9781760573454
Paperback: 164 pages
Dimensions: 6.14 x 0.35 x 9.21 inches
Language: eng

"M'lle Modiste": A Comic Opera in Two Acts

Henry Blossom & Victor Herbert

ISBN: 9781760574567
Paperback: 224 pages
Dimensions: 6.14 x 0.47 x 9.21 inches
Language: eng

A Doctor of the Old School

Ian Maclaren

www.triestepublishing.com

You May Also Like

Clarendon Press Series. Easy Passages for Translation into Latin

John Young Sargent

ISBN: 9781760579708
Paperback: 176 pages
Dimensions: 6.0 x 0.38 x 9.0 inches
Language: eng

A Free Translation, in Verse, of the "Inferno" of Dante, with a Preliminary Discourse and Notes

Dante Alighieri & Bruce Whyte

ISBN: 9781760574666
Paperback: 210 pages
Dimensions: 6.14 x 0.44 x 9.21 inches
Language: eng

www.triestepublishing.com

You May Also Like

Shut Your Mouth and Save Your Life

George Catlin

ISBN: 9781760570491
Paperback: 118 pages
Dimensions: 6.14 x 0.25 x 9.21 inches
Language: eng

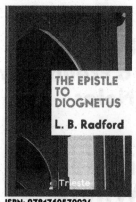

The Epistle to Diognetus

L. B. Radford

ISBN: 9781760570934
Paperback: 106 pages
Dimensions: 6.14 x 0.22 x 9.21 inches
Language: eng

www.triestepublishing.com

You May Also Like

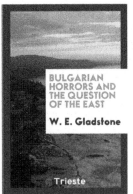

Bulgarian horrors and the question of the East

W. E. Gladstone

ISBN: 9781760571146
Paperback: 46 pages
Dimensions: 6.14 x 0.09 x 9.21 inches
Language: eng

Snow-bound: A Winter Idyl

John Greenleaf Whittier

ISBN: 9781760571528
Paperback: 64 pages
Dimensions: 5.5 x 0.13 x 8.25 inches
Language: eng

Find more of our titles on our website. We have a selection of thousands of titles that will interest you. Please visit

www.triestepublishing.com

Lightning Source UK Ltd.
Milton Keynes UK
UKOW01f0018120917
309021UK00004B/456/P